letters to a young mathematician

By the Same Author

Concepts of Modern Mathematics
Game, Set, and Math
Does God Play Dice?
Another Fine Math You've Got Me Into
Fearful Symmetry
Nature's Numbers
From Here to Infinity
The Magical Maze
Life's Other Secret
Flatterland
What Shape Is a Snowflake?
The Annotated Flatland
Math Hysteria
The Mayor of Uglyville's Dilemma

with Jack Cohen
The Collapse of Chaos
Figments of Reality
What Does a Martian Look Like?
Wheelers (science fiction)
Heaven (science fiction)

with Terry Pratchett and Jack Cohen
The Science of Discworld
The Science of Discworld II: The Globe
The Science of Discworld III: Darwin's Watch

C.1

Ian Stewart

letters to a young
mathematician

A Member of the Perseus Books Group
New York

Published by Basic Books,
A Member of the Perseus Books Group

Books published by Basic Books are available at special discounts for bulk purchases in the United States by corporations, institutions, and other organizations. For more information, please contact the Special Markets Department at the Perseus Books Group, 11 Cambridge Center, Cambridge MA 02142, or call (617) 252–5298 or (800) 255–1514, or e-mail special.markets@perseusbooks.com.

Library of Congress Cataloging-in-Publication Data
Stewart, Ian.
    Letters to a young mathematician / Ian Stewart.
        p.   cm.
    Includes bibliographical references.
    ISBN–13: 978–0–465–08231–5 (alk. paper)
    ISBN–10: 0–465–08231–9 (alk. paper)
        1. Mathematics—Miscellanea.   I. Title.
QA99.S84   2006
510—dc22

2005030384

06 07 08 / 10 9 8 7 6 5 4 3 2 1

In memory of
Marjorie Kathleen ("Madge") Stewart
4.2.1914–17.12.2001
and
Arthur Reginald ("Nick") Stewart
2.3.1914–23.8.2004
without whom I would not have been anything,
let alone a mathematician.

Contents

Contents

◼ Preface

"It is a melancholy experience for a professional mathematician to find himself writing about mathematics." So the great English mathematician Godfrey Harold Hardy, of the University of Cambridge, opened his 1940 classic *A Mathematician's Apology*.

Attitudes change. No longer do mathematicians believe that they owe the world an apology. And many are now convinced that writing *about* mathematics is at least as valuable as writing mathematics, by which Hardy meant new mathematics, new research, new theorems. In fact, many of us feel that it is pointless for mathematicians to invent new theorems unless the public gets to hear of them. Not the details, of course, but the general nature of the enterprise. In particular, that new mathematics is constantly being created, and what it is used for.

The world has changed, too, since Hardy's time. A typical day for Hardy consisted in a maximum of four hours of intensive thought about research problems; the rest of the day was then occupied watching the game of cricket, his great nonmathematical passion, and reading the newspapers. He must have fitted in some time for the occasional research student as well, but he was reticent about personal matters. A typical day for the modern academic is ten or twelve hours long, with teaching commitments, research grants to pursue, research to be carried out, and liberal doses of pointless bureaucracy to get in the way of anything creative.

Hardy was typical of a certain kind of English academic. He set himself high but narrow standards. He valued his chosen field for its own internal elegance and logic, not for its external uses. He was proud that none of his work could have any possible use in warfare, a position with which most of us can sympathize, especially bearing in mind that his book was published in the opening years of World War II.

He would be disappointed in the extreme to be resurrected today and to learn that on the contrary, his beloved theory of numbers plays an essential role in the mathematical theory of cryptography, with evident military uses. The movie *Enigma* paints a romanticized view of the period when this connection first began to emerge, in the vital wartime work of the code breakers at Bletchley Park. Prominent among them was the tragic figure of Alan Turing—pure mathematician, applied mathematician, and pioneering computer scientist—who committed suicide because he was persecuted for being a homosexual, a sexual orientation that was then illegal and considered shameful. Social mores change, too.

Hardy's classic little gem sheds a great deal of light on how academic mathematicians viewed themselves and their subject in 1940. It contains important lessons for any would-be young mathematician, but some of these are obscured by the book's outdated attitudes, such as its default assumption that mathematics is strictly a male preserve. It is still worth reading, but only if its opinions are seen in their historical context, and it is not assumed that they all remain valid today.

Letters to a Young Mathematician is my attempt to bring some parts of *A Mathematician's Apology* up to date, namely, those parts that might influence the decisions of a young person contemplating a degree in mathematics and a possible career in the subject. The letters, addressed to "Meg," follow her career in roughly chronological order, from high school through to a tenured position in a university. They discuss a variety of topics, ranging from basic career decisions to the working philosophy of professional mathematicians and the nature of their subject. The intention is not merely to offer practical advice, but to give an inside view of the mathematical enterprise, and to explain what it is really like to be a mathematician.

As a result, many of the issues discussed will also appeal to a more general audience, the one for which

Hardy wrote: anyone who is interested in mathematics and its relation to human society. What is mathematics? What is it good for? How can you learn it? How can you teach it? Is it a solitary activity or can it be done in groups? How does the mathematical mind work? And where is it all going?

I would never have thought of writing *Letters to a Young Mathematician* were it not for Basic Books, and the wonderful mentoring series to which this book belongs. The book benefited from the advice of my editor, Bill Frucht, who made sure that I confined my ramblings to the topic at hand and made them accessible. The main intended readership is the "young mathematician" of the title, or their parents, relatives, friends ... but the book should appeal to anyone who is interested in what it is like to become, and be, a mathematician, even if they have no such ambitions themselves.

Ian Stewart
Coventry, September 2005

1

■ Why Do Math?

Dear Meg,

As you probably anticipated, I was very glad to hear you're thinking of studying mathematics, not least because it means all those weeks you spent reading and rereading *A Wrinkle in Time* a few summers ago were not wasted, nor all the hours I spent explaining tesseracts and higher dimensions to you. Rather than deal with your questions in the order you asked them, let me take the most practical one first: does anyone besides me actually make a living doing math?

The answer is different from what most people think. My home university did a survey of its alumni a few years back, and they discovered that out of all the various degree subjects, the one that led to the highest average income was . . . mathematics. Mind you, that was before they opened the new medical school, but it demolishes one myth: that mathematics can't lead to a well-paying job.

The truth is that we encounter mathematicians everywhere, every day, but we hardly ever know it. Past students of mine have managed breweries, started their own electronics companies, designed automobiles, written software for computers, and traded futures on the stock market. It simply doesn't occur to us that our bank manager might have a degree in math, or that the people who invent or manufacture DVDs and MP3 players employ large numbers of mathematicians, or that the technology that transmits those stunning pictures of the moons of Jupiter relies heavily on math. We know that our doctor has a medical degree, and our lawyer has a law degree, because those are specific, well-defined professions that require equally specific training. But you don't find brass plaques on buildings advertising a licensed mathematician within, who, for a large fee, will solve any math problems that you need help with.

Our society consumes an awful lot of math, but it all happens behind the scenes. The reason is straightforward: that's where it belongs. When you drive a car, you don't want to have to worry about all the complicated mechanical things that make it work; you want to get in and drive away. Sure, it helps you to be a better driver if you're aware of the basics of car mechanics, but even that is not essential. It's the same with math. You want your car navigation system to give you directions *without* your having to do the math yourself. You want your phone to work without your having to understand signal processing and error-correcting codes.

Some of us, however, need to know how to do the math, or none of these wonders could function. It would be great if the rest of us were aware of just how strongly we rely on mathematics in our daily lives; the problem with putting math so far behind the scenes is that many people have no idea it's there at all.

I sometimes think that the best way to change the public attitude to math would be to stick a red label on everything that uses mathematics. "Math inside." There would be a label on every computer, of course, and I suppose if we were to take the idea literally, we ought to slap one on every math teacher. But we should also place a red math sticker on every airline ticket, every telephone, every car, every airplane, every traffic light, every vegetable . . .

Vegetable?

Yes. The days when farmers simply planted what their fathers had planted, and their fathers before them, are long gone. Virtually any plant you can buy is the outcome of a long and complicated commercial breeding program. The whole topic of "experimental design," in the mathematical sense, was invented in the early 1900s to provide a systematic way to assess new breeds of plants, not to mention the newer methods of genetic modification.

Wait. Isn't this biology?

Biology, sure. But math, too. Genetics was one of the first parts of biology to go mathematical. The Human Genome Project succeeded because of a lot of clever work by biologists, but a vital feature of the entire project was the development of powerful mathematical methods to analyze the experimental results and reconstruct accurate genetic sequences from very fragmentary data.

So, vegetables get a red sticker. Just about everything there is gets a red sticker.

You go to movies? Do you like the special effects? *Star Wars*, *Lord of the Rings*? Mathematics. The first full-length computer-animated movie, *Toy Story*, led to the publication of about twenty research papers on math. "Computer graphics" isn't just computers making pictures; it's the mathematical methods that make those pictures look realistic. To do that, you need three-dimensional geometry, the mathematics of light, "in-betweening" to interpolate a smooth series of images between a start and a finish, and lots more. "Interpolation" is a mathematical idea. Computers are clever engineering, but they don't do anything useful without a lot of clever math. Red sticker.

And then, of course, there's the Internet. If anything makes use of math, it's the Internet. The main search engine at the moment, Google, was founded on a mathematical method for working out which web pages are most likely to contain the information required by a user. It's based on matrix algebra, probability theory, and the combinatorics of networks.

But the math of the Internet is much more fundamental than that. The telephone network relies on math. It's not like the old days when switchboard operators literally connected calls by plugging phone lines in by hand. Today those lines have to carry millions of messages at once. There are so many of us, all wanting to talk to our friends or send faxes or access the Internet, that we have to share the phone lines and the suboceanic cables and the satellite relays, or the network wouldn't be able to carry all that traffic. So each conversation is broken up into thousands upon thousands of short pieces, and only one piece in a hundred is actually transmitted. At the other end, the missing ninety-nine pieces are restored by filling in the gaps as smoothly as possible (it works because the samples, though short, are very frequent, so that the sounds you make when you speak change much more slowly than the interval between samples). Oh, and the entire signal is coded so that any transmission errors can not only be detected, they can be put right at the receiving end.

Modern communications systems simply would not work without a huge quantity of math. Coding theory, Fourier analysis, signal processing . . .

Anyway, you go onto the Internet to get a plane ticket, book your flight and turn up at the airport, hop on the plane, and away you go. The plane flies because the engineers who designed it used the mathematics of fluid flow, aerodynamics, to make sure it would stay up. It navigates using a global positioning system (GPS), a system of satellites whose signals, analyzed mathematically, can tell you where you are to within a few feet. The flights have to be scheduled so that each plane is in the right place when it is next needed, rather than somewhere on the far side of the globe, and that, again, requires yet other areas of math.

And so, Meg, my dear, it goes. You asked me whether mathematicians are all shut away in universities, or whether some of them do work related to real life. Your entire life bobs like a small boat on a vast ocean of mathematics.

But hardly anyone notices. Hiding the math away makes us all feel comfortable, but it devalues mathematics. That is a shame. It makes people think that math isn't useful, that it doesn't matter, that it's just intellectual games without any true significance. Which is why I'd like to see those red stickers. In fact, the best reason not to use them is that most of the planet would be covered with them.

Your third question was the most important, and the saddest. You asked me whether you would have to give up your sense of beauty to study mathematics, whether everything would become just numbers and equations to you, laws and formulas. Rest assured, Meg, I don't blame you for asking this, since it's unfortunately a very common idea, but it couldn't be more wrong. It's exactly the opposite of the truth.

What math does for me is this: It makes me aware of the world I inhabit in an entirely new way. It opens my eyes to nature's laws and patterns. It offers an entirely new experience of beauty.

When I see a rainbow, for instance, I don't just see a bright, multicolored arc across the sky. I don't just see the effect of raindrops on sunlight, splitting the white light from the sun into its constituent colors. I still find rainbows beautiful and inspiring, but I appreciate that there's more to a rainbow than mere refraction of light. The colors are, so to speak, a red (and blue and green) herring. What require explanation are the shape and the brightness. Why is a rainbow a circular arc? Why is the light from the rainbow so bright?

You may not have thought about those questions. You know that a rainbow appears when sunlight is refracted by tiny droplets of water, with each color of light being diverted through a slightly different angle and bouncing back from the raindrops to meet the observing eye. But if that's all there is to a rainbow, why don't the billions of differently colored light rays from billions of raindrops just overlap and smear out?

The answer lies in the geometry of the rainbow. When the light bounces around inside a raindrop, the spherical shape of the drop causes the light to emerge with a very strong focus along a particular direction.

Each drop in effect emits a bright cone of light, or, rather, each color of light forms its own cone, and the angle of the cone is slightly different for each color. When we look at a rainbow, our eyes detect only the cones that come from raindrops lying in particular directions, and for each color, those directions form a circle in the sky. So we see lots of concentric circles, one for each color.

The rainbow that you see and the rainbow that I see are created by different raindrops. Our eyes are in different places, so we detect different cones, produced by different drops.

Rainbows are personal.

Some people think that this kind of understanding "spoils" the emotional experience. I think this is rubbish. It demonstrates a depressing sort of aesthetic complacency. People who make such statements often like to pretend they are poetic types, wide open to the world's wonders, but in fact they suffer from a serious lack of curiosity: they refuse to believe the world is more wonderful than their own limited imaginations. Nature is always deeper, richer, and more interesting than you thought, and mathematics gives you a very powerful way to appreciate this. The ability to *understand* is one of the most important differences between human beings and other animals, and we should value it. Lots of animals emote, but as far as we know, only humans think rationally. I'd say that my understanding of the geometry of the rainbow adds a new dimension to its beauty. It doesn't take anything away from the emotional experience.

The rainbow is just one example. I also look at animals differently, because I'm aware of the mathematical patterns that underlie their movements. When I look at a crystal, I am aware of the beauties of its atomic lattice as well as the charm of its colors. I see mathematics in waves and sand dunes, in the rising and the setting of the sun, in raindrops splashing in a puddle, even in birds sitting on telephone cables. And I'm aware—dimly, as if looking out over a foggy ocean—of the infinity of things we *don't* know about these everyday wonders.

Then there's the inner beauty of mathematics, which should not be underrated. Math done "for its own sake" can be exquisitely beautiful and elegant. Not the "sums" we all do at school; as individuals those are mostly ugly and formless, although the general principles that govern them have their own kind of beauty. It's the ideas, the generalities, the sudden flashes of insight, the realization that trying to trisect an angle with straightedge and compass is like trying to prove that 3 is an even number, that it makes perfect sense that you can't construct a regular seven-sided polygon but you can construct one with seventeen sides, that there is no way to untie an overhand knot, and why some infinities are bigger than others whereas some that ought to be bigger are actually equal, that the *only* square number (other than 1, if you want to be picky) that is the sum of consecutive squares, $1 + 4 + 9 + \cdots$, is the number 4900.

You, Meg, have the potential to become an accomplished mathematician. You have a logical mind and also an inquiring one. You're not convinced by vague arguments; you want to see the details and check them out for yourself. You don't just want to know how to make things work, you want to know *why* they work. And your letter made me hope that you'll come to see mathematics as I see it, as something fascinating and beautiful, a way of seeing the world that is like no other.

I hope this sets the scene for you.

Yours,
Ian

2

■ **How I Almost Became a Lawyer**

Dear Meg,

You ask how I got into mathematics. As with anyone, it was a combination of talent (there's no point in being modest), encouragement, and the right sort of accident, or more accurately, being rescued from the wrong sort of accident.

I was good at math from the start, but when I was seven, I very nearly got put off the subject for life. There was a math test, and we were supposed to subtract the numbers, but I did the same as the previous week and added them. So I got a zero and was put in the lower section of the class. Because the other kids in that section were hopeless at math, we didn't do anything interesting. I wasn't being challenged, and I got bored.

I was saved by two things: a broken bone and my mom.

One of the other kids pushed me over in the playground as part of a game, and I broke my collarbone. I was out of school for five weeks, so Mom decided to make good use of the time. She borrowed the arithmetic book from the school, and we did some remedial work. Because I couldn't write—my right hand was in a sling—I dictated the numbers, and she wrote them in the exercise book.

My mother was rather sensitive about schooling. Her own education had been pretty much ruined by the mistaken good intentions of a well-meaning but unimaginative school inspector. Because she was quick on the uptake, she was advanced rapidly through the grades until, by the age of eight, she was in with a class of ten-year-olds. The school inspector came by one day, observed the class, and asked the intelligent little girl who was answering all the questions, "How old are you, my dear?" On being told "eight," he informed the school principal that the bright little girl must stay in the same class for three years in a row, until the other kids her age caught up with her. He wasn't trying to hold her back academically; he was worried that she was out of her depth socially. But repeating the same lessons three years in a row killed off my mother's interest in school; all she learned was how to goof off.

Later she worked out what had happened, but by then it was too late. She wanted to be an English teacher, but she failed her chemistry exam. In those days, in the United Kingdom, failing just one subject, even one that was totally irrelevant to the subject you wanted to teach, meant that you could not train as a teacher.

My mother was determined that nothing similar should happen to me. She *knew* I was clever; she'd taught me to read when I was three. After we had done 400 math problems and I'd gotten 396 of them right, she took the exercise book into school, showed it to the head teacher, and demanded that I be moved into the top section of the math class.

When my collarbone healed and I went back to school, I was ten weeks ahead of the rest of the class in math. We'd overdone it a bit. Fortunately, I didn't suffer too much while the class caught up.

My teacher wasn't a bad teacher. In fact, he was a very kind man. But he lacked the imagination to realize that he'd put me in the wrong section, and that his mistake was going to damage my education. I'd gotten a zero on the test because I was careless, not because I didn't understand the material. If he'd simply told me to read the questions carefully, I'd have gotten the point.

I was lucky then, thanks mostly to my mother's good sense and willingness to fight for me. But I also owe a debt to my schoolmate for putting me in the hospital. He'd done it quite unintentionally—we were all shoving each other around—but it saved my mathematical bacon.

After that I had several really brilliant math teachers. And those, let me tell you, are rare. There was one named W. E. Beck (we nicknamed him "spider") whose Friday math test was a long-standing institution. Those were not easy tests. They were graded out of twenty points, and as the weeks passed, each kid's grades were added up. The kids who were good at math were desperate to come in first for the year; the others were just desperate. I'm not sure it was acceptable educational practice—in fact I'm sure it wasn't—but the competitive element was good for me and a few buddies.

One of Beck's rules was that if you missed a test, even if you were sick, you got zero. No excuses. So those of us who were in the running needed to make every point count. We knew we needed a cushion, since you weren't safe unless you were ahead by more than twenty points. So you absolutely did not lose points by making silly mistakes. You read every question, made sure you'd done what was asked, checked everything, and then you checked it again.

Later, when I was sixteen, I had a math teacher named Gordon Radford. Normally he was lucky to get one boy who was really talented at math, but in my class there were six of us. So he spent all of his free periods teaching us extra math, outside the syllabus. During the regular math lessons he told us to sit at the back and do our homework; not just math, *any* homework. And to shut up. Those lessons weren't for us; we had to give the others a chance.

Mr. Radford opened my eyes to what math was really like: diverse, creative, full of novelty and originality. And he did one more crucial thing for me.

In those days, there was a public entrance exam called a State Scholarship that provided funding to go to college. You still needed to be offered a place, but a State Scholarship was a big step in the right direction. In the last year that State Scholarships were to be offered, I and two friends were a year too young to take the exam. Mr. Radford had to persuade the headmaster to put us in for the exam one year early, something the headmaster never did.

One morning when my two friends and I arrived at school, Mr. Radford told us we would be joining the class one year ahead of us to take a "mock exam" for the State Scholarship in math. A practice run. The older kids had done a year's more math and had been practicing for weeks; we had five minutes' warning. I came in first, and my friends were second and third.

So the headmaster had no choice but to let us take the exam for the State Scholarship. After all, he was letting the older kids take it, and we had proved we were better prepared for it than they were.

All three of us were awarded State Scholarships.

At that point Mr. Radford got in touch with David Epstein, whom he had taught some years before and who had become a mathematician at Cambridge University, along with Oxford, the United Kingdom's leading university, especially renowned for its math.

"What do I do with this boy?" Gordon asked.

"Send him to us," said David.

So I went to study math at Cambridge, the home of Isaac Newton, Bertrand Russell, and Ludwig Wittgenstein (along with many lesser lights), and I never looked back.

Some careers seem to accumulate people who might easily have preferred to do something else. You will run into people who tell you that they practice law as a day job but they are really novelists or playwrights or jazz trombonists. Other people can't settle on something, or they see their careers in more purely practical terms, and they drift into human resources management or advertising sales. Which is not to say that these people are not dedicated or fulfilled in what they do, but few of them consider their work a *calling*.

No one drifts into being a mathematician. On the contrary, it's a pursuit from which even the talented are too easily turned away. If I hadn't broken my collarbone, if Mr. Beck hadn't fostered all-out competition among his students, if there hadn't been an unusually large group of strong students for Mr. Radford to promote— and if he hadn't done it so aggressively—instead of writing you today I might be telling your parents how to save more on their tax return. And perhaps no one, least of all me, would suspect that things could have turned out differently.

In short, Meg, you should not expect your teachers to look at you once and simply *see*, in a brief glance, how bright you are. You should not expect them to unerringly spot your talents and know where they might lead you. Some will, and you will be grateful to them for the rest of your life. But others, sadly, can't tell, or don't much care, or are caught up in their own worries and resentments. Then again, the ones who stand in awe of your gifts are not the ones from whom you will ultimately learn the most. The best teachers will occasionally, perhaps more than occasionally, make you feel a bit stupid.

3

■ The Breadth of Mathematics

Dear Meg,

It's not hard to see, in your question, a sense of—I don't know—anticipated boredom, or perhaps some worry about what you've let yourself in for. It's all reasonably interesting now, but, as you say, "Is this all there is?" You're reading Shakespeare, Dickens, and T. S. Eliot in your English class, and you can reasonably assume that while this is of course only a tiny sample of the world's great writing, there is not some higher level of English literature whose existence has not been disclosed to you. So you naturally wonder, by analogy, whether the math you're learning in high school is what mathematics *is*. Does anything happen at higher levels besides bigger numbers and harder calculations?

What you've seen so far is not really the main event.

Mathematicians do not spend most of their time doing numerical calculations, even though calculations are sometimes essential to making progress. They do not occupy themselves with grinding out symbolic formulas, but formulas can nonetheless be indispensable. The school math you are learning is mainly some basic tricks of the trade, and how to use them in very simple contexts. If we were talking woodwork, it's like learning to use a hammer to drive a nail, or a saw to cut wood to size. You never see a lathe or an electric drill, you do not learn how to build a chair, and you absolutely do not learn how to design and build an item of furniture no one has thought of before.

Not that a hammer and saw aren't useful. You can't make a chair if you don't know how to cut the wood to the correct size. But you should not assume that because that's all you ever did at school, it's all carpenters ever do.

An awful lot of what is now called "mathematics" at school is really arithmetic: various notations for numbers, and methods for adding, subtracting, multiplying, and dividing them. As you get older, you are shown other bits of the toolkit: elementary algebra, trigonometry, coordinate geometry, maybe a little calculus. If your syllabus was "modernized" in the 1960s or 1970s, you may get two-by-two matrices and tiny bits of group theory. "Modern" is a strange word to use here: it means between one and two hundred years old, as opposed to the two-hundred-plus-year-old math that formed the bulk of the older syllabus.

Unfortunately, it's almost impossible to progress to the more interesting regions of the subject if you don't know how to do sums and get them right, how to solve basic equations, or what an ellipse is. The highest levels of every human activity demand a solid grasp of the basics; think of tennis, or playing the violin. Mathematics happens to require rather a lot of basic knowledge and technique.

At university you will encounter a much broader conception of mathematics. In addition to the familiar numbers, there will be complex numbers, where minus one has a square root. Things far more important than numbers will appear, such as functions: rules that assign, for any chosen number, some specific other number. "Square," "cosine," "cube root," those are all functions. You won't just solve simultaneous equations in two unknowns; you will understand the solutions of simultaneous equations in any number of unknowns, when they exist at all, which they sometimes don't. (Try solving $x + y = 1$, $2x + 2y = 3$.) You may learn how the great mathematicians of the Renaissance solved cubic and quartic equations (involving cubes and fourth powers of the unknown), not just quadratics; if so, you will probably find out why such methods fail for quintic equations (fifth powers). You will see why this becomes almost obvious if you ignore the numerical values of the equations' solutions and instead think about their symmetries, and why it is arguably more important to understand the symmetries of equations than to be able to solve them.

You will find out how to formalize the concept of symmetry in abstract terms, which is what group theory is. You will discover that Euclid's geometry is not the only one possible, and move on to topology, where circles and triangles become indistinguishable. You will have your intuition challenged by Möbius bands, which are surfaces with only one side, and fractals, which are shapes so complex that they have a fractional number of dimensions. You will learn methods to solve differential equations, and eventually you will appreciate that most of them cannot be solved by those methods; then you will learn how they can still be understood and used, even when you cannot write down their solutions. You will find out why every number can be resolved uniquely into prime factors, be puzzled by the apparent lack of patterns in primes despite their statistical regularities, and baffled by open questions like the Riemann hypothesis. You will meet different sizes of infinity, discover the real reasons why π is important, and prove that knots exist. You will belatedly realize just how abstract your subject has become, how far removed from mere numbers, and then numbers will bite you on the ankle, reemerging as key ideas.

You will learn why tops wobble and how that affects ice ages; you will comprehend Newton's proof that planets' orbits are elliptical, and find out why they aren't *perfectly* elliptical, opening up the Pandora's box of chaotic dynamics. Your eyes will be opened to the vast range of uses of math, from the statistics of plant breeding to the orbital dynamics of space probes, from

Google to GPS, from ocean waves to the stability of bridges, from the graphics in *Lord of the Rings* to antennas for mobile phones.

You will come to feel just how much of our world would be impossible without math.

And when you survey this glorious diversity, you will wonder what makes it all the same thing: why are such disparate types of ideas all called mathematics? You will have gone from asking "Is this all there is?" to being slightly amazed that there can be so much. By then, just as you can recognize a chair but can't define one in a manner that permits no exceptions, you will find that you can recognize mathematics when you see it, but you still can't define it.

Which is as it should be. Definitions pin things down, they limit the prospects for creativity and diversity. A definition, implicitly, attempts to reduce all possible variations of a concept to a single pithy phrase. Math, like anything still under development, always has the potential to surprise.

Schools—not just yours, Meg, but around the world—are so preoccupied with teaching sums that they do a poor job of preparing students to answer (or even ask) the far more interesting and difficult question of what mathematics *is*. And even though definitions are too limiting, we can still try to capture the flavor of our subject, using something that the human brain is unusually good at: metaphor. Our brains are not like computers, working systematically and logically. They are metaphor machines that leap to creative conclusions and belatedly shore them up with logical narratives. So, when I tell you that one of my favorite "definitions" of math is Lynn Arthur Steen's phrase "the science of significant form," you may feel that I've made a useful stab at the question, metaphorically speaking.

What I like about Steen's metaphor is that it captures some crucial features. Above all, it is open-ended; it does not attempt to specify what kind of form should be considered significant, or what "form" or "significant" are even supposed to mean. I also like the word "science," because math shares far more with the sciences than it does with the arts. It has the same reliance on stringent testing, except that in science this is done through experiments, whereas math employs proofs. It has the same character of operating within closely specified constraints: you can't just make it up as you go along. Here I part company with the postmodernists, who assert that everything (except, apparently, postmodernism) is merely a social convention. Science, they tell us, consists only of opinions that happen to be held by a lot of scientists. Sometimes this *is* the case—the prevalent belief that the human sperm count is falling is probably an example—but mostly it is not. There is no question that science has a social side, but it also has the reality check of experiment. Even postmodernists must always enter a room through the door, not through the wall.

There is a famous book called *What Is Mathematics?* written by Richard Courant and Herbert Robbins. As with most books whose titles are questions, the question is never quite answered. Yet the authors say some very wise things. Their prologue begins, "Mathematics as an expression of the human mind reflects the active will, the contemplative reason, and the desire for aesthetic perfection." It goes on to tell us that "All mathematical development has its psychological roots in more or less practical requirements. But once started under the pressure of necessary applications, it inevitably gains momentum in itself and transcends the confines of immediate utility." And it ends like this: "Fortunately, creative minds forget dogmatic philosophical beliefs whenever adherence to them would impede constructive achievement. For scholars and layman alike it is not philosophy but active experience in mathematics itself that alone can answer the question: What is mathematics?" Or, as my friend David Tall often says, "Math is not a spectator sport."

Some mathematicians are more interested in the philosophy of their subject than others, and among to-day's prominent philosophers of mathematics we find Reuben Hersh. He observed that Courant and Robbins answered their question "by *showing* what mathematics is, not by *telling* what it is. After devouring the book with wonder and delight, I was still left asking, 'But what is mathematics really?'" So Hersh wrote a book with that title, offering what he said was an unconventional answer.

Traditionally, there have been two main schools of mathematical philosophy: Platonism and formalism. Platonists believe that in some (slightly mystical) way, mathematical objects exist. They are "out there" in some abstract realm. This realm is not imaginary, however, because imagination is a human characteristic. It is *real*, in a nonphysical sense. The mathematician's circle, with its infinitely thin circumference and a radius that remains constant to infinitely many decimal places, cannot take physical form. If you draw it in sand, as Archimedes did, its boundary is too thick and its radius too variable. Your drawing is only an approximation of the mathematical, Platonic circle. Inscribe it on a platinum slab with a diamond-tipped needle—the same difficulties still arise.

In what sense, then, does a mathematical circle *exist*? And if it doesn't, how can it be useful? Platonists tell us that the mathematical circle is an ideal, not realized in this world but nevertheless having a reality that is independent of human minds.

Formalists find such statements fuzzy and meaningless. The first major formalist was David Hilbert, and he tried to put the whole of mathematics on a sound logical basis by effectively treating it as a meaningless game played with symbols. A statement like 2 + 2 = 4 was not, from this point of view, to be interpreted in terms of, say, putting two sheep in a pen with two others and thus having four sheep. It was the outcome of a game played with the symbols 2, 4, +, and =. But the game must be played according to an explicit list of absolutely rigid rules.

Philosophically, formalism died when Kurt Gödel proved, to Hilbert's initial fury, that no formal theory can capture the whole of arithmetic *and* be proved logically consistent. There will always be mathematical statements that remain outside Hilbert's game: neither provable nor disprovable. Any such statement can be added to the axioms for arithmetic without creating any inconsistency. The negation of such a statement has the same feature. So we can deem such a statement to be true, or we can deem it false, and Hilbert's game can be played either way. In particular, the idea that arithmetic is so basic and natural that it has to be unique is wrong.

Most working mathematicians have ignored this, just as they have ignored the apparent mysticism of the Platonist view, probably because the interesting questions in math are those that can either be proved or disproved. When you are doing math, it *feels* as though what you are working on is real. You can almost pick things up and turn them around, squash them and stroke them and pull them to pieces. On the other hand, you often make progress by forgetting what it all means and focusing solely on how the symbols dance. So the working philosophy of most mathematicians is a mostly unexamined Platonist–Formalist hybrid.

That's fine if all you want is to *do* mathematics. As Hersh says, "Mathematics comes first, then philosophizing about it, not the other way round." But if, like Hersh, you still wonder whether there might be a better way to describe that working philosophy, it all comes back to that same basic question of what mathematics is.

Hersh's answer is what he calls the humanist philosophy. Mathematics is "A human activity, a social phenomenon, part of human culture, historically evolved, and intelligible only in a social context." This is a description, not a definition, since it does not specify the content of that activity. The description may sound a bit postmodern, but it is made more intelligent than postmodernism by Hersh's awareness that the social conventions that govern the activities of those human minds are subject to stringent *non*social constraints, namely, that everything must fit together logically. Even if mathematicians got together and agreed that π equals 3, it wouldn't. Nothing would make sense.

A mathematical circle, then, is something more than a shared delusion. It is a concept endowed with extremely specific features; it "exists" in the sense that human minds can deduce other properties from those features, with the crucial caveat that if two minds investigate the same question, they cannot, by correct reasoning, come up with contradictory answers.

That's why it feels as if math is "out there." Finding the answer to an open question feels like discovery, not invention. Math is a product of human minds but not bendable to human will. Exploring it is like exploring a new tract of country; you may not know what is around the next bend in the river, but you don't get to choose. You can only wait and find out. But the mathematical countryside does not come into existence until you explore it.

When two members of the Arts Faculty argue, they may find it impossible to reach a resolution. When two mathematicians argue—and they do, often in a highly emotional and aggressive way—suddenly one will stop, and say, "I'm sorry, you're quite right, now I see my mistake." And they will go off and have lunch together, the best of friends.

I agree with Hersh, pretty much. If you feel that the humanist description of math is a bit woolly, that this type of "shared social construct" is a rarity, Hersh offers some examples that might change your mind. One is money. The entire world runs on money, but what is it? It is not pieces of paper or disks of metal; those can be printed or minted anew, or handed into a bank and destroyed. It is not numbers in a computer: if the computer blew up, you would still be entitled to your money. Money is a shared social construct. It has value because we all agree it has value.

Again, there are strong constraints. If you tell your bank manager that your account contains more than his computer says it does, he does not respond, "No problem, it's just a social construct, here's an extra ten million dollars. Have a nice day."

It is tempting to think that even if we consider math to be a shared social construct, it has a kind of logical inevitability, that any intelligent mind would come up with the *same* math. When the *Pioneer* and *Voyager* spacecraft were sent off into space, they carried coded messages from humanity to any alien race that might one day encounter them. *Pioneer* bore a plaque with a diagram of the hydrogen atom, a map of nearby pulsars to show where our sun is located, line drawings of a naked man and woman standing in front of a sketch of the spacecraft, for scale, and a schematic picture of the solar system to show which planet we inhabit. The two *Voyager* craft carried records with sounds, music, and scientific images.

Would an alien recipient be able to decode those messages? Would a picture like o–o, two circles joined by a line, *really* look like the hydrogen atom to them? What if their version of atomic theory relied on quantum wave functions instead of primitive "particle" images, which even our own physicists tell us are wildly inaccurate? Would the aliens understand line drawings, given that humans from tribes that have never encountered such things fail to do so? Would they consider pulsars significant?

In most discussions about such questions, one eventually hears it argued that even if they grasped nothing else, any intelligent alien would be able to comprehend simple mathematical patterns, and the rest can be built from there. The unstated assumption is that math is somehow *universal*. Aliens would count 1, 2, 3, . . . just as we do. They would surely see the implied pattern in diagrams like * ** *** **** .

I'm not convinced. I've been reading *Diamond Dogs*, by Alistair Reynolds, a novella about an alien construct, a bizarre and terrifying tower, through whose rooms you progress by solving puzzles. If you get the answer wrong, you die, horribly. Reynolds's story is powerful, but there is an underlying assumption that aliens would set mathematical puzzles akin to those that a human would set. Indeed, the alien math is *too* close to human; it includes topology and an area of mathematical physics known as Kaluza–Klein theory. You are as likely to arrive on the fifth planet of Proxima Centauri and find a Wal-Mart. I know that narrative constraints demand that the math should *look* like math to the reader, but even so, it doesn't work for me.

I think human math is more closely linked to our particular physiology, experiences, and psychological preferences than we imagine. It is parochial, not universal. Geometry's points and lines may seem the natural basis for a theory of shape, but they are also the features into which our visual system happens to dissect the world. An alien visual system might find light and shade primary, or motion and stasis, or frequency of vibration. An alien brain might find smell, or embarrassment, but not shape, to be fundamental to its perception of the world. And while discrete numbers like 1, 2, 3, seem universal to us, they trace back to our tendency to assemble similar things, such as sheep, and consider them property: has one of *my* sheep been stolen? Arithmetic seems to have originated through two things: the timing of the seasons and commerce. But what of the blimp creatures of distant Poseidon, a hypothetical gas giant like Jupiter, whose world is a constant flux of turbulent winds, and who have no sense of individual ownership? Before they could count up to three, whatever they were counting would have blown away on the ammonia breeze. They would, however, have a far better understanding than we do of the math of turbulent fluid flow.

I *think* it is still credible that where blimp math and ours made contact, they would be logically consistent with each other. They could be distant regions of the same landscape. But even that might depend on which type of logic you use.

The belief that there is *one* mathematics—ours—is a Platonist belief. It's possible that "the" ideal forms are "out there," but also that "out there" might comprise more than one abstract realm, and that ideal forms need not be unique. Hersh's humanism becomes Poseidonian blimpism: their math would be a social construct shared by *their* society. If they had a society. If they didn't—if different blimps did not communicate—could they have any conception of mathematics at all? Just as we can't imagine a mathematics not founded on the counting numbers, we can't imagine an "intelligent" species whose members don't communicate with each other. But the fact that we can't imagine something is no proof that it doesn't exist.

But I am drifting off the topic. What is mathematics? In despair, some have proposed the definition "Mathematics is what mathematicians do." And what are mathematicians? "People who do mathematics." This argument is almost Platonic in its perfect circularity. But let me ask a similar question. What is a businessman? Someone who does business? Not quite. It is someone who *sees opportunities* for doing business when others might miss them.

A mathematician is someone who sees opportunities for doing mathematics.

I'm pretty sure that's right, and it pins down an important difference between mathematicians and everyone else. What is mathematics? It is the shared social construct created by people who are aware of certain opportunities, and we call those people mathematicians. The logic is still slightly circular, but mathematicians can always recognize a fellow spirit. Find out what that fellow spirit does; it will be one more aspect of our shared social construct.

Welcome to the club.

4

■ **Hasn't It All Been Done?**

Dear Meg,

In your last letter you asked me about the extent to which mathematics at university can go beyond what you have already done at school. No one wants to spend three or four years going over the same ideas, even if they are studied in greater depth. Now, looking ahead, you are also right to worry about the scope that exists for creating new mathematics. If others have already explored such a huge territory, how can you ever find your way to the frontier? Is there even any frontier left?

For once, my task is simple. I can set you at ease on both counts. If anything, you should worry about the exact opposite: that people are creating too much new mathematics, and that the scope for new research is so gigantic that it will be difficult to decide where to start or in which direction to proceed. Math is not a robotic way of replacing thought by rigid ritual. It is the most creative activity on the planet.

These statements will be news to many people, possibly including some of your teachers. It always astonishes me that so many people seem to believe that mathematics is limited to what they were taught at school, so that basically "it's all been done." Even more astonishing is the assumption that because "the answers are all in the back of the book," there is no scope for creativity, and no questions remain unanswered. Why do so many people think that their school textbook contains every possible question?

This failure of imagination would amount to deplorable ignorance, were it not for two factors that together go a long way to explain it.

The first is that many students quickly come to dislike mathematics as they pass through the school system. They find it rigid, boring, repetitive, and, worst of all, difficult. Answers are either right or wrong, and no amount of clever verbal jousting with the teacher can convert a wrong answer into a correct one. Mathematics is a very unforgiving subject. Having developed this negative attitude, the last thing the student wants to hear is that there is more mathematics, going beyond the already daunting contents of the set text. Most people *want* all the answers to be at the back of the book, because otherwise they can't look them up.

Dame Kathleen Ollerenshaw, one of Britain's most distinguished mathematicians and educators, who continues to do research at the age of ninety, makes exactly this point in her autobiography *To Talk of Many Things*. (Do read it, Meg; it's inspirational, and very wise.) "When I told a teenage friend that I was doing mathematical research, her reply was, 'Why do that? We have enough mathematics to cope with already—we don't want any more.'"

The assumptions behind that statement bear examination, but I content myself with just one. Why did Kathleen's friend assume that any newly invented mathematics would automatically appear in school texts? Again we encounter the same belief, that the math you are taught at school is the entire universe of mathematics. But no one thinks that about physics, or chemistry, or biology, or even French or economics. We all know that what we are taught at school is just a tiny part of what is currently known.

I sometimes wish schools would go back to using words like "arithmetic" to describe the content of "math" courses. Calling them "mathematics" debases the currency of mathematical thought; it's a bit like using the term "composing" to describe routine exercises in playing musical scales. However, I lack the power to change the name, and if in fact the name were changed, the main effect would be to *decrease* public recognition of mathematics. For most people, the only time they are aware of their life and mathematics intersecting is at school.

As I wrote in my first letter, this does not imply that mathematics has no relevance to our daily lives.

But the profound influence of our subject on human existence takes place behind the scenes and therefore passes unnoticed.

A second reason why few students ever realize that there is mathematics outside the textbook is that no one ever tells them that.

I don't blame the teachers. Math is actually very important, but because it genuinely is difficult, nearly all of the teaching slots are occupied with making sure that students learn how to solve certain types of problem and get the answers right. There isn't time to tell them about the history of the subject, about its connections with our culture and society, about the huge quantity of new mathematics that is created every year, or about the unsolved questions, big and little, that litter the mathematical landscape.

Meg, the *World Directory of Mathematicians* contains fifty-five thousand names and addresses. These people don't just sit on their hands. They teach, and most of them do research. The journal *Mathematical Reviews* appears twelve times per year, and the 2004 issues totaled 10,586 pages. But this journal does not consist of research papers; it consists of brief *summaries* of research papers. Each page summarizes, on average, about five papers, so for that year the summaries covered about fifty thousand actual papers. The average size of a paper is perhaps twenty pages—roughly a million pages of new mathematics every year!

Kathleen's friend would have been horrified.

Many teachers are aware of this, but they have a good reason not to say much about it. If your students are having problems remembering how to solve quadratic equations, the wise teacher will stay well clear of cubic equations, which are even more difficult. When the issue in class is finding solutions of simultaneous equations that possess solutions, it would be demoralizing and confusing to inform the students that many sets of simultaneous equations have no solutions at all, and others have infinitely many. A process of self-censorship sets in. In order to avoid damaging the students' confidence, the texts do not ask questions that the methods being taught cannot answer. So, insidiously, we absorb the lesson that every mathematical question has an answer.

It's not true.

Our teaching of mathematics revolves around a fundamental conflict. Rightly or wrongly, students are required to master a series of mathematical concepts and techniques, and anything that might divert them from doing so is deemed unnecessary. Putting mathematics into its cultural context, explaining what it has done for humanity, telling the story of its historical development, or pointing out the wealth of unsolved problems or even the existence of topics that do not make it into school textbooks leaves less time to prepare for the exam. So mostly these things aren't discussed. Some teachers—my Mr. Radford was an example—find time to fit them in anyway. Ellen and Robert Kaplan, an American husband-and-wife team with a refreshing approach to mathematics education, have started a series of "math circles," where young children are encouraged to think about mathematics in an atmosphere that could not be more different from that of a classroom.

Their success shows that we need to set aside more time in the syllabus for such activities. But since math already occupies a substantial part of teaching time, people who teach other subjects might object. So the conflict may well remain unresolved.

Now let me explain a wonderful thing: the more mathematics you learn, the more opportunities you will find for asking new questions. As our knowledge of mathematics grows, so do the opportunities for fresh discoveries. This may sound unlikely, but it is a natural consequence of how new mathematical ideas build on older ones.

When you study any subject, the rate at which you can understand new material tends to accelerate the more you already know. You've learned the rules of the game, you've gotten good at playing it, so learning the next level is easier. At least it would be, except that at higher levels you set yourself higher standards. Math is like that. To perhaps an extreme degree, it builds new concepts on top of old ones. If math were a building, it would resemble a pyramid erected upside down. Built on a narrow base, the structure would tower into the clouds, each floor larger than the one below.

The taller the building becomes, the more space there is to build more.

That's perhaps a little too simple a description. There would be funny little excrescences protruding all over the place, twisting and turning; decorations like minarets and domes and gargoyles; stairways and secret passageways unexpectedly connecting distant rooms; diving boards suspended over dizzying voids. But the inverted pyramid would dominate.

All subjects are like that to some extent, but their pyramids do not widen so rapidly, and new buildings are often put up beside existing ones. These subjects resemble cities, and if you don't like the building you are in, you can always move to another one and start afresh.

Mathematics is *all one thing*, and moving house is not an option.

Because school math is heavily biased toward numbers, many people think that math comprises only numbers, that mathematical research must consist in inventing *new* numbers. But of course there aren't any, are there? If there were, someone would already have invented them. But this belief is a failure of imagination, even when it comes to numbers.

Most schoolwork on numbers is arithmetical. Add 473 to 982. Divide 16 by 4. A lot is about notation: fractions like 7/5, decimals like 1.4, recurring decimals like $0.3333\ldots$, or more obstreperous numbers like π, whose decimal digits go on forever without any repetitive pattern.

How do we know that about π? Not by listing every digit, or by listing lots of them and failing to observe any repetition. By proving it, indirectly. The first such proof was published in 1770 by Johann Lambert, and it is based not on geometry but on calculus. It occupies about a page and is mostly a calculation. The trick is not the calculation but figuring out which calculation to do.

A few more inventive topics also appear at school level, such as prime numbers, which cannot be obtained by multiplying two smaller (whole) numbers together. But pretty much everything students are exposed to boils down to buttons you could push on a pocket calculator.

The higher floors of the mathematical anti-pyramid do not look like this at all. They support concepts, ideas, and processes. They address questions very different from "add these two numbers," such as, "Why do the digits of π not repeat?" The floors that *do* deal with numbers rapidly get to extremely difficult questions, which often appear deceptively straightforward.

For instance, you will be aware that a triangle with sides 3, 4, and 5 units long has a right angle; allegedly the ancient Egyptians used a string divided into such lengths by knots to survey the building site for the pyramids. I am skeptical about the practical use of the 3–4–5 triangle, because string can stretch and I doubt that the measurements can be carried out to the required accuracy, but the Egyptians probably knew the triangle's properties. Certainly the ancient Babylonians did.

Pythagoras's theorem—one of the few theorems mentioned at school that bears the name of its (traditional) discoverer—tells us that the squares of the two shorter sides add up to the square of the longer one: $3^2 + 4^2 = 5^2$. There are infinitely many such "Pythagorean triangles," and ancient Greek mathematicians already knew how to find them all. Pierre de Fermat, a seventeenth-century French lawyer whose hobby was mathematics, asked the kind of imaginative question (not *very* imaginative; you don't have to go far beyond what is already known to encounter yawning gaps in human knowledge) that creates new mathematics. We know about sums of two squares making squares, but can you do it with cubes? Can two cubes add up to a cube? Or two fourth powers to a fourth power? Fermat could not discover any solutions. He found an elegant proof that it can't be done with fourth powers. In his copy of an ancient Greek number theory text, he stated that he had a proof that it can't be done generally—that there are no solutions in whole numbers to the equation $x^n + y^n = z^n$, where n is greater than 2—but "this margin is too small to contain it."

Leave aside the question of the utility of such mathematics; applications are important too, but right now we're talking about creativity and imagination. Take *too* "practically minded" an attitude and you stifle true creativity, to everyone's detriment. Fermat's last theorem, as the problem came to be known, turned out to be very deep and very hard. It is unlikely that Fermat's proof, if it existed, was correct. If it was, no one else has ever thought of it, not even now, when we know Fermat was right. Generations of mathematicians attacked the problem and came away with nothing. A few chipped the odd corner off it; they proved that it couldn't be done with fifth powers, say, or seventh powers. Only in 1994, after a hiatus of 350 years, was the theorem proved, by Andrew Wiles; his proof was published the following year. You probably remember a TV documentary about it.

Wiles's methods were revolutionary, and much too difficult even for a university course at undergraduate or introductory graduate level. His proof is very clever and very beautiful, incorporating results and ideas from dozens of other experts. A breakthrough of the highest order.

The TV program was very moving. Many viewers burst into tears.

The proof of Fermat's last theorem leaps right over the undergraduate syllabus. It is too advanced for the courses you will take. But you will certainly take more elementary courses in number theory, proving theorems like "every positive integer is a sum of at most four squares." You may elect to study algebraic number theory, where you will see how the great mathematicians of past eras chipped pieces off Fermat's last theorem, and understand how the whole of abstract algebra emerged from that process. This is a new world that goes almost totally unnoticed by the great majority of humanity.

Nearly everyone makes use of number theory every day, if only because it forms the basis of Internet security codes and the data-compression methods employed by cable and satellite television. We don't need to be able to *do* number theory to watch TV (otherwise ratings of many shows would be way down), but if nobody knew any number theory, crooks would be helping themselves to our bank accounts, and we'd be stuck with three channels. So the general area of math in which Fermat's last theorem lives is undoubtedly useful.

The theorem itself, though, is unlikely to be of much use. Very few practical problems rest on adding two big powers together to get another such power. (Though I am told that at least one problem in physics does depend on this.) Wiles's new methods, on the other hand, have opened up significant new connections between hitherto separate areas of our subject. Those methods will surely turn out to be important one day, very likely in fundamental physics, which is today's biggest consumer of deep, abstract mathematical concepts and techniques.

Questions like Fermat's last theorem are not important because we need to know the answer. In the end it probably doesn't matter that the theorem was proved true rather than false. They are important because our efforts to find the answer reveal major gaps in our understanding of mathematics. What counts is not the answer itself but knowing how to get it. It can only go in the back of the book when someone has worked out what it is.

The further we push out the boundaries of mathematics, the bigger the boundary itself becomes. There is no danger that we will ever run out of new problems to solve.

5

■ Surrounded by Math

Dear Meg,

I'm not surprised that you're "both excited and a little bit intimidated," as you put it, by your imminent move to university. Let me commend your good intuition on both counts. You'll find the competition tougher, the pace faster, the work harder, and the content far more interesting. You'll be thrilled by your teachers (some of them) and the ideas they lead you to discover, and daunted that so many of your classmates seem to get there ahead of you. For the first six months you'll wonder why the school ever let you in. (After that you'll wonder how some of the others were let in.)

You asked me to tell you something inspirational. Nothing technical, just something to hold on to when the going gets tough.

Very well.

Like many mathematicians, I get my inspiration from nature. Nature may not look very mathematical; you don't see sums written on the trees. But math is not about sums, not really. It's about patterns and why they occur. Nature's patterns are both beautiful and inexhaustible.

I'm in Houston, Texas, on a research visit, and I'm surrounded by math.

Houston is a huge, sprawling city. Flat as a pancake. It used to be a swamp, and when there's a heavy thunderstorm, it tries to revert to its natural condition. Close by the apartment complex where my wife and I always stay when we visit, there is a concrete-lined canal that diverts a lot of the runoff from the rain. It doesn't always divert quite enough; a few years ago the nearby freeway was thirty feet under water, and the ground floor of the apartment complex was flooded. But it helps. It's called Braes Bayou, and there are paths along both sides of it. Avril and I like to go for walks along the bayou; the concrete sides are not exactly pretty, but they're prettier than the surrounding streets and parking lots, and there's quite a lot of wildlife: catfish in the river, egrets preying on the fish, lots of birds.

As I walk along Braes Bayou, surrounded by wildlife, I realize that I am also surrounded by math.

For instance . . .

Roads cross the bayou at regular intervals, and the phone lines cross there too, and birds perch on the phone lines. From a distance they look like sheet music, fat little blobs on rows of horizontal lines. There seem to be special places they like to perch, and it's not at all clear to me why, but one thing stands out. If a lot of birds are perching on a wire, they end up *evenly spaced*.

That's a mathematical pattern, and I think there's a mathematical explanation. I don't think the birds "know" they ought to space themselves out evenly. But each bird has its own "personal space," and if another birds gets too close, it will sidle along the wire to leave a bit more room, unless there's another bird crowding it from the other side.

When there are just a few birds, they end up randomly spaced. But when there are a lot, they get pushed close together. As each one sidles along to make itself feel more comfortable, the "population pressure" evens them out. Birds at the edge of denser regions get pushed into less densely populated regions. And since the birds are all of the same species (usually they're pigeons), they all have much the same idea of what their personal space should be. So they space themselves evenly.

Not *exactly* evenly, of course. That would be a Platonic ideal. As such, it helps us to comprehend a more messy reality.

You could do the math on this problem if you wanted to. Write down some simple rules for how birds move when the neighbors get too close, plonk them down at random, run the rules, and watch the spacing evolve. But there's an analogy with a common physical system, where that math has already been done, and the analogy tells you what to expect.

It's a *bird crystal*.

The same process that makes birds space themselves regularly makes the atoms in a solid object line up to form a repetitive lattice. The atoms also have a "personal space": they repel each other if they're too close together. In a solid, the atoms are forced to pack fairly tightly, but as they adjust their personal spaces, they arrange themselves in an elegant crystal lattice.

The bird lattice is one-dimensional, since they're sitting on a wire. A one-dimensional lattice consists of equally spaced points. When there are just a few birds, arranged at random and not subject to population pressure, it's not a crystal, it's a gas.

This isn't just a vague analogy. The *same* mathematical process that creates a regular crystal of salt or calcite also creates my "bird crystal."

And that's not the only math that you can find in Braes Bayou.

A lot of people walk their dogs along the paths. If you watch a walking dog, you quickly notice how rhythmic its movement is. Not when it stops to sniff at a tree or another dog, mind you; it's rhythmic only when the dog is just bumbling happily along without a thought in its head. Tail wagging, tongue lolling, feet hitting the ground in a careless doggy dance.

What do the feet do?

When the dog is walking, there's a characteristic pattern. Left *rear*, left *front*, right *rear*, right *front*. The footfalls are equally spaced in time, like musical notes, four beats to the bar.

If the dog speeds up, its gait changes to a trot. Now diagonal pairs of legs—left rear and right front, then the other two—hit the ground together, in an alternating pattern of two beats to the bar. If two people walked one behind the other, exactly out of step, and you put them inside a cow costume, the cow would be trotting.

The dog is math incarnate. The subject of which it is an unwitting example is known as gait analysis; it has important applications in medicine: humans often have problems moving their legs properly, especially in infancy or old age, and an analysis of how they move can reveal the nature of the problem and maybe help cure it. Another application is to robotics: robots with legs can move in terrain that doesn't suit robots with wheels, such as the inside of a nuclear power station, an army firing range, or the surface of Mars. If we can understand legged locomotion well enough, we can engineer reliable robots to decommission old power stations, locate unexploded shells and mines, and explore distant planets. Right now, we're still using wheels for Mars rovers because that design is reliable, but the rovers are limited in where they can go. We're not decommissioning nuclear power stations at all. But the U.S. Army does use legged robots for some tidying-up tasks on firing ranges.

If we learn to reinvent the leg, all that will change.

Egrets standing in the shallows with that characteristic alert posture, long beaks poised, muscles tensed, are hunting catfish. Together they form a miniature ecology, a predator–prey system. Ecology's connection with mathematics goes all the way back to Leonardo of Pisa, also known as Fibonacci, who wrote about a rather simple model of the growth of rabbits in 1202, in his *Liber Abaci*. To be fair, the book is really about the Hindu-Arabic number system, the forerunner of today's ten-symbol notation for numbers, and the rabbit model is mainly there as an exercise in arithmetic. Most of the other exercises are currency transactions; it was a very practical book.

More-serious ecological models arose in the 1920s, when the Italian mathematician Vito Volterra was trying to understand a curious effect that had been observed by Adriatic fishermen. During World War I, when the amount of fishing was reduced, the numbers of food fish didn't seem to increase, but the population of sharks and rays did.

Volterra wondered why a reduction in fishing benefited the predators more than it benefited the prey. To find out, he devised a mathematical model, based on the sizes of the shark and food-fish populations and how each affected the other. He discovered that instead of settling down to steady values, populations underwent repetitive cycles: large populations became smaller but then increased, over and over again. The shark population peaked sometime after the food-fish population did.

You don't need numbers to understand why. With a moderate number of sharks, the food fish can reproduce faster than they are eaten, so their population soars. This provides more food for the sharks, so their population also begins to climb; but they reproduce more slowly, so there is a delay. As the sharks increase in number, they eat more food fish, and eventually there are so many sharks that the food-fish population starts to decline. Now the food fish cannot support so many sharks, so the shark numbers also drop, again with a delay. With the shark population reduced, the food fish can once more increase . . . and so it goes.

The math makes this story crystal clear (within the assumptions built into the model) and also lets us work out how the average population sizes behave over a complete cycle, something the verbal argument can't handle. Volterra's calculations showed that a reduced level of fishing decreases the average number of food fish over a cycle but increases the average number of sharks. Which is just what happened during World War I.

All of the examples I've told you about so far involve "advanced" mathematics. But simple math can also be illuminating. I am reminded of one of the many stories mathematicians tell each other after all the nonmathematicians leave the room. A mathematician at a famous university went to look around the new auditorium, and when she got there, she found the dean of the faculty staring at the ceiling and muttering to himself, ". . . forty-five, forty-six, forty-seven . . ." Naturally she interrupted the count to find out what it was for. "I'm counting the lights," said the dean. The mathematician looked up at the perfect rectangular array of lights and said, "That's easy, there are . . . twelve that way, and . . . eight that way. Twelve eights are ninety-six." "No, no," said the dean impatiently. "I want the *exact* number."

Even when it comes to something as simple as counting, we mathematicians see the world differently from other folk.

6

◼ How Mathematicians Think

Dear Meg,

I would say you've lucked out. If you're hearing about people like Newton, Leibniz, Fourier, and others, it means your freshman calculus teacher has a sense of the history of his subject; and your question "How did they think of these things?" suggests that he's teaching calculus not as a set of divine revelations (which is how it's too often done) but as real problems that were solved by real people.

But you're right, too, that the answer "Well, they were geniuses" isn't really adequate. Let me see if I can dig a little deeper. The general form of your question—which is a very important one—is "How do mathematicians think?"

You might reasonably conclude from looking at textbooks that all mathematical thought is symbolic. The words are there to separate the symbols and explain what they signify; the core of the description is heavily symbolic. True, some areas of mathematics make use of pictures, but those are either rough guides to intuition or visual representations of the results of calculations.

There is a wonderful book about mathematical creation, *The Psychology of Invention in the Mathematical Field*, by Jacques Hadamard. It was first published in 1945, and it's still in print and extremely relevant today. I recommend you pick up a copy. Hadamard makes two main points. The first is that most mathematical thinking begins with vague visual images and is only later formalized with symbols. About ninety percent of mathematicians, he tells us, think that way. The other ten percent stick to symbols the entire time. The second is that ideas in mathematics seem to arise in three stages.

First, it is necessary to carry out quite a lot of conscious work on a problem, trying to understand it, exploring ways to approach it, working through examples in the hope of finding some useful general features. Typically, this stage bogs down in a state of hopeless confusion, as the real difficulty of the problem emerges.

At this point it helps to stop thinking about the problem and do something else: dig in the garden, write lecture notes, start work on another problem. This gives the subconscious mind a chance to mull over the original problem and try to sort out the confused mess that your conscious efforts have turned it into. If your subconscious is successful, even if all it manages is to get part way, it will "tap you on the shoulder" and alert you to its conclusions. This is the big "aha!" moment, when the little lightbulb over your head suddenly switches on.

Finally, there is another conscious stage of writing everything down formally, checking the details, and organizing it so that you can publish it and other mathematicians can read it. The traditions of scientific publication (and of textbook writing) require that the "aha!" moment be concealed, and the discovery presented as a purely rational deduction from known premises.

Henri Poincaré, probably my favorite among the great mathematicians, was unusually aware of his own thought processes and lectured about them to psychologists. He called the first stage "preparation," the second "incubation followed by illumination," and the third "verification." He laid particular emphasis on the role of the subconscious, and it is worth quoting one famous section of his essay *Mathematical Creation*:

> *For fifteen days I strove to prove that there could not be any functions like those I have since called Fuchsian functions. I was then very ignorant; every day I seated myself at my table, stayed an hour or two, tried a great number of combinations and reached no results. One evening, contrary to my custom, I drank black coffee and could not sleep. Ideas rose in crowds; I felt them collide until pairs interlocked, so to speak, making a stable combination. By the next morning I had established the existence of a class of Fuchsian*

*functions, those which come from the hypergeometric
series; I had only to write out the results, which took
but a few hours.*

This was but one of several occasions on which Poincaré felt that he was "present at his own unconscious work."

A recent experience of my own also fits Poincaré's three-stage model, though I did not have the feeling that I was observing my own subconscious. A few years ago, I was working with my long-term collaborator Marty Golubitsky on the dynamics of networks. By "network" I mean a set of dynamical systems that are "coupled together," with some influencing the behavior of others. The systems themselves are the nodes of the network—think of them as blobs—and two nodes are joined by an arrow if one of them (at the tail end) influences the other (at the head end). For example, each node might be a nerve cell in some organism, and the arrows might be connections along which signals pass from one cell to another.

Marty and I were particularly interested in two aspects of these networks: synchrony and phase relations. Two nodes are synchronous if the systems they represent do exactly the same thing at the same moment. That trotting dog synchronizes diagonally opposite legs: when the front left foot hits the ground, so does the back right. Phase relations are similar, but with a time lag. The dog's front right foot (which is similarly synchronized with its back left foot) hits the ground half a cycle later than the front left foot. This is a half-period phase shift.

We knew that synchrony and phase shifts are common in symmetric networks. In fact, we had worked out the only plausible symmetric network that could explain all of the standard gaits of four-legged animals. And we'd sort of assumed, because we couldn't think of any other reason, that symmetry was also necessary for synchrony and phase shifts to occur.

Then Marty's postdoc Marcus Pivato invented a very curious network that had synchrony and phase shifts but no symmetry. It had sixteen nodes, which synchronized in clusters of four, and each cluster was separated from one of the others by a phase shift of one quarter of a period. The network was almost symmetric at first sight, but when you looked closely you could see that the apparent symmetry was imperfect.

To us, Marcus's example made absolutely no sense. But there was no question that his calculations were correct. We could check them, and we did, and they worked. But we were left with a nagging feeling that we didn't really understand *why* they worked. They involved a kind of coincidence, which definitely happened, but "shouldn't have."

While Marty and Marcus worked on other topics, I worried about Marcus's example. I went to Poland for a conference and to give some lectures, and for the whole of that week I doodled networks on notepads. I doodled all the way from Warsaw to Krakow on the train, and two days later I doodled all the way back. I felt I was close to some kind of breakthrough, but I found it impossible to write down what it might be.

Tired and fed up, I abandoned the topic, shoved the doodles into a filing cabinet, and occupied my time elsewhere. Then one morning I woke up with a strange feeling that I should dig out the file and take another look at the doodles. Within minutes I had noticed that all the doodles that did what I wanted had a common feature, one that I'd totally missed when I was doodling them. Not only that; all of the doodles that didn't do what I wanted lacked that feature. At that moment I "knew" what the answer to the puzzle was, and I could even write it down symbolically. It was neat, tidy, and very simple.

The trouble with that kind of knowledge, as my biologist friend Jack Cohen often says, is that it feels just as certain when you're wrong. There is no substitute for proof. But now, because I knew *what* to prove and had a fair idea of why it was true, that final stage didn't take very long. It was blindingly obvious how to prove that the feature that I had observed in my doodles was *sufficient* to make happen everything I thought should happen. Proving that it was also necessary was trickier, but not greatly so. There were several relatively obvious lines of attack, and the second or third worked.

Problem solved.

This description fits Poincaré's scenario so perfectly that I worry that I have embroidered the tale and rearranged it to make it fit. But I'm pretty sure that it really did happen the way I've just told you

What was the key insight? I've just looked through my notes from the Warsaw–Krakow train, and they are full of networks whose nodes have been colored. Red, blue, green . . . At some stage I had decided to color the nodes so that synchronous nodes got the same color. Using the colors, I could spot hidden regularities in the networks, and those regularities were what made Marcus's example work. The regularities weren't symmetries, not in the technical sense used by mathematicians, but they had a similar effect.

Why had I been coloring the networks? Because the colors made it easy to pick out the synchronous clusters. I had colored dozens of networks and never noticed what the colors were trying to tell me. The answer had been staring me in the face. But only when I stopped working on the problem did my subconscious have the freedom to sort it out.

It took a week or two to turn this insight into formal mathematics. But the visual thinking—the colors—came first, and my subconscious had to grapple with the problem before I was consciously aware of the answer. Only then did I start to reason symbolically.

There's more to the tale. Once the formal system was sorted out, I noticed a deeper idea, which underlay the whole thing. The similarities between colored cells formed a natural algebraic structure. In our previous work on symmetric systems we had put a similar structure in from the very start, because all mathematicians know how to formalize symmetries. The concept concerned is called a group. But Marcus's network has no symmetry, so groups won't help. The natural algebraic structure that replaces the symmetry group in my colored diagrams is something less well known, called a "groupoid."

Pure mathematicians have been studying groupoids for years, for their own private reasons. Suddenly I realized that these esoteric structures are intimately connected with synchrony and phase shifts in networks of dynamical systems. It's one of the best examples, among the topics that I've been involved with, of the mysterious process that turns pure math into applications.

Once you understand a problem, many aspects of it suddenly become much simpler. As mathematicians the world over say, everything is either impossible or trivial. We immediately found lots of simpler examples than Marcus's. The simplest has just two nodes and two arrows.

Research is an ongoing activity, and I think we have to go further than Hadamard and Poincaré to understand the process of invention, or discovery, in math. Their three-stage description applies to a single "inventive step" or "advance in understanding." Solving most research problems involves a whole series of such steps.

In fact, any step may break down into a series of sub-steps, and those substeps may also break down in a similar manner. So instead of a single three-stage process, we get a complicated network of such processes. Hadamard and Poincaré described a basic tactic of mathematical thought, but research is more like a strategic battle. The mathematician's strategy employs that tactic over and over again, on different levels and in different ways.

How do you learn to become a strategist? You take a leaf from the generals' book. Study the tactics and strategies of the great practitioners of the past and present. Observe, analyze, learn, and internalize. And one day, Meg—closer than you might think—other mathematicians will be learning from *you*.

7

■ How to Learn Math

Dear Meg,

By now, you've surely noticed that the quality of teaching in a college setting varies widely. This is because, for the most part, your professors and their teaching assistants are not hired, kept on, or promoted based on their ability to teach. They are there to do research, whereas teaching, while necessary and important for any number of reasons, is decidedly secondary for many of them. Many of your professors will be thrilling lecturers and devoted mentors; others, you'll find, will be considerably less thrilling and devoted. You'll have to find a way to succeed even with teachers whose greatest talents are not necessarily on display in the classroom.

I once had a lecturer who, I was convinced, had discovered a way to make time stand still. My classmates disagreed with this thesis but felt that his sleep-inducing powers must surely have military uses.

The vast amounts that have been written about teaching math might give the impression that all of the difficulties encountered by math students are caused by teachers, and it is always the teacher's responsibility to sort out the student's problems. This is, of course, one of the things teachers are paid to do, but there is some onus on the student as well. You need to understand how to learn.

Like all teaching, math instruction is rather artificial. The world is complicated and messy, with lots of loose ends, and the teacher's job is to impose order on the confusion, to convert a chaotic set of episodes into a coherent narrative. So your learning will be divided into specific modules, or courses, and each course will have a carefully specified syllabus and a text. In some settings, such as some American public schools, the syllabus will specify exactly which pages of the text, and which problems, are to be tackled on a given day. In other countries and at more advanced levels, the lecturer has more of a free hand to pick his or her own path through the material, and the lecture notes take the place of a textbook.

Because the lectures progress through set topics, one step at a time, it is easy for students to think that this is how to learn the material. It is not a bad idea to work systematically through the book, but there are other tactics you can use if you get stuck.

Many students believe that if you get stuck, you should stop. Go back, read the offending passage again;

repeat until light dawns—either in your mind or outside the library window.

This is almost always fatal. I always tell my students that the first thing to do is read on. Remember that you encountered a difficulty, don't try to pretend that all is sweetness and light, but continue. Often the next sentence, or the next paragraph, will resolve your problem.

Here is an example, from my book *The Foundations of Mathematics*, written with David Tall. On page 16, introducing the topic of real numbers, we remark that "the Greeks discovered that there exist lines whose lengths, in theory, cannot be measured exactly by a *rational* number."

One might easily grind to a halt here—what does "measured by" mean? It hasn't been defined yet, and—oh help—it's not in the index. And how did the Greeks discover this fact anyway? Am I supposed to know it from a previous course? From this course? Did I miss a lecture? The previous pages of the book offer no assistance, however many times you reread them. You could spend hours getting nowhere.

So don't. Read on. The next few sentences explain how Pythagoras's theorem leads to a line whose length is the square root of two, and state that there is no rational number m/n such that $(m/n)^2 = 2$. This is then proved, cleverly using the fact that every whole number can be expressed as a product of primes in only one way. The result is summarized as "no rational number can have square 2, and hence that the hypotenuse of the given triangle does not have rational length."

By now, everything has probably slotted into place. "Measured by" presumably means "has a length equal to." The Greek reasoning alluded to in such an offhand manner is no doubt the argument using Pythagoras's theorem; it helps to know that Pythagoras was Greek. And you should be able to spot that "the square root of two is not rational" is equivalent to "no rational number can have square 2."

Mystery solved.

If you are *still* stuck after plowing valiantly ahead in search of enlightenment, now is the time to go to your class tutor or the lecturer and ask for assistance. By trying to sort the problem out for yourself, you will have set your mind in action, and thus you are much more likely to understand the teacher's explanation. It's much like Poincaré's "incubation" stage of research. Which, with fair weather and a following wind, leads to illumination.

There is another possibility, but it's one where help from the teacher is probably essential. Even so, you can try to prepare the ground. Whenever you get stuck on a piece of mathematics, it usually happens because you do not properly understand some *other* piece of mathematics, which is being used without explicit mention on the assumption that you can handle it easily. Remember the upside-down pyramid of mathematical knowledge? You may have forgotten what a rational number is, or what

Pythagoras proved, or how square roots relate to squares. Or you may be wondering why the uniqueness property of prime factorization is relevant. If so, you don't need help to understand the proof that the square root of two is irrational; you need help rehearsing rational numbers, prime factors, or basic geometry.

It takes a certain insight into your own thought processes as well as a certain discipline to pinpoint exactly what you don't understand and relate it to your immediate difficulty. Your tutors know about such things and will be on the lookout for them. It is, however, a very useful trick to master for *yourself*, if you can.

To sum up: If you think you are stuck, begin by plowing ahead regardless, in the hope of gaining enlightenment, but remember where you got stuck, in case this doesn't work. If it doesn't, return to the sticking point and backtrack until you reach something you are confident you understand. Then try moving ahead again.

This process is very similar to a general method for solving a maze, which computer scientists call "depth-first search." If possible, move deeper into the maze. If you get stuck, backtrack to the first point where there is an alternative path, and follow that. Never go over the exact same path twice. This algorithm will get you safely through any maze. Its learning analogue does not come with such a strong guarantee, but it's still a very good tactic.

As a student I took this method to extremes. My usual method for reading a mathematics text was to thumb through it until I spotted something interesting, then work backward until I had tracked down everything I needed to read the interesting bit. I don't really recommend this to everyone, but it does show that there are alternatives to starting at page 1 and continuing in sequence until you get to page 250.

Let me urge upon you another useful trick. It may sound like a huge amount of extra work, but I assure you it will pay dividends.

Read around your subject.

Do not read only the assigned text. Books are expensive, but universities have libraries. Find some books on the same topic or similar topics. Read them, but in a fairly casual way. Skip anything that looks too hard or too boring. Concentrate on whatever catches your attention. It's amazing how often you will read something that turns out to be helpful next week, or next year.

The summer before I went off to Cambridge to study math, I read dozens of books in this easygoing way. One of them, I remember, was about "vectors," which the author defined as "quantities that have both magnitude and direction." At the time this made very little sense to me, but I liked the elegant formulas and simple diagrams with lots of arrows, and I skimmed through it more than once. I then forgot it. In the opening lecture on vectors,

Ian Stewart

suddenly it all clicked into place. I understood exactly what the author had been trying to tell me, *before* the lecturer got that far. All those formulas seemed obvious: I knew why they were true.

I can only assume that my subconscious had been stirred up, just as Poincaré claimed, and during the intervening period, it had created some order out of my desultory wanderings through that book on vectors. It was just waiting for a few simple clues before it could form a coherent picture.

When I say "read around your subject," I don't mean just the technical material. Read Eric Temple Bell's *Men of Mathematics*, still a cracking read even if some of the stories are invented and women are almost invisible. Sample the great works of the past; James Newman's *The World of Mathematics* is a four-volume set of fascinating writings about math from ancient Egypt through to relativity. There has been a spate of popular math books in recent years, on the Riemann hypothesis, the four color theorem, π, infinity, mathematical crackpots, how the human brain thinks mathematical thoughts, fuzzy logic, Fibonacci numbers. There are even books on the applications of mathematics, such as D'Arcy Thompson's classic *On Growth and Form*, about mathematical patterns in living creatures. It may be outmoded in biological terms—it was written long before the structure of DNA was found—but its overall message remains as valid as ever.

Such books will broaden your appreciation of what math is, what it can be used for, and how its sits in human culture. There will likely be no questions about any of these topics on your exams. But awareness of these issues will make you a better mathematician, able to grasp the essentials of any new topic more confidently.

There are also some specific techniques that will improve your learning skills. The great American mathematical educator George Pólya put a lot of them into his classic *How to Solve It*. He took the view that the only way to understand math properly is the hands-on method: tackling problems and solving them. He was right. But you can't learn this way if you get stuck on every problem you try. So your teachers will set you a carefully chosen sequence of problems, starting with routine calculations and leading up to more challenging questions.

Pólya offers many tricks for boosting your problem-solving abilities. He describes them far better than I can, but here is a sample. If the problem seems baffling, try to recast it in a simpler form. Look for a good example and try your ideas out on the example; later, you can generalize to the original setting. For instance, if the problem is about prime numbers, try it on 7, 13, or 47. Try working backward from the conclusion: what steps must we take to get there? Try several examples and look for common patterns; if you find one, try to prove that it must *always* happen.

As you remarked in your letter, Meg, one of the main differences between high school and college is that in college the students are treated much more like adults. This means that to a much greater extent, it's sink or swim: pass, fail, or find another major. There is plenty of help available for the asking, but that too takes more initiative than it did in high school. No one is likely to take you by the hand and say, "It looks like you're having trouble."

On the other hand, the rewards for self-sufficiency are much greater. Your high school was mainly grateful if you were not a problem requiring some sort of extra attention, and unless you were extremely lucky, the most it could offer an exceptional student (beyond the grades certifying him or her to move on) was an extracurricular club and perhaps an award or two. In a university you will encounter real scholars who are on the lookout for young people capable of doing real mathematics, and they are just waiting for you to stand out, if you can.

8

■ Fear of Proofs

Dear Meg,

You're quite correct: One of the biggest differences between school math and university math is proof. At school we learn *how* to solve equations or find the area of a triangle; at university we learn *why* those methods work, and prove that they do. Mathematicians are obsessed with the idea of "proof." And, yes, it does put a lot of people off. I call them proofophobes. Mathematicians, in contrast, are proofophiles: no matter how much circumstantial evidence there may be in favor of some mathematical statement, the true mathematician is not satisfied until the statement is *proved*. In full logical rigor, with everything made precise and unambiguous.

There's a good reason for this. A proof provides a cast-iron guarantee that some idea is correct. No amount of experimental evidence can substitute for that.

Let's take a look at a proof and see how it differs from other forms of evidence. I don't want to use anything that involves technical math, because that will obscure the underlying ideas. My favorite nontechnical proof is the SHIP–DOCK theorem, which is about those word games in which you have to change one word into another by a sequence of moves: CAT, COT, COG, DOG. At each step, you are allowed to change (but not move) exactly one letter, and the result must be a valid word (as determined by, say, Webster's).

Solving this word puzzle isn't particularly hard: for instance,

SHIP
SHOP
SHOT
SLOT
SOOT
LOOT
LOOK
LOCK
DOCK

There are plenty of other solutions. But I'm not after a solution as such, or even several: I'm interested in something that applies to *every* solution. Namely, at some stage, there must be a word that contains two vowels. Like SOOT (and LOOT and LOOK) in this particular answer. Here I mean *exactly* two vowels, no more and no less.

To avoid objections, let me make it clear what "vowel" means here. One thorny problem is the letter Y. In YARD the letter Y is a consonant, but in WILY it is a vowel. Similarly, the W in CWMS acts as a vowel: "cwm" is Welsh, and refers to a geological formation for which there seems to be no English word, although "corrie" (Scottish) and "cirque" (French) are alternatives. We need to be very careful about letters that sometimes act as vowels but on other occasions are consonants. In fact, the safest way to avoid the kinds of words that all Scrabble players love is to throw away Webster's and redefine "vowel" and "word" in a more limited sense. For the purposes of this discussion, a "vowel" will mean one of the letters A, E, I, O, U, and a "word" will be required to contain *at least one* of those five letters. Alternatively, we can require Y and W *always* to count as vowels, even when they are being used as consonants. What we can't do, in this context, is allow letters to be sometimes vowels, sometimes consonants. I'll come back to that later.

It's not a question of what the correct convention is in linguistics; I'm setting up a temporary convention for a specific mathematical purpose. Sometimes in math the best way to make progress is to introduce simplifications, and that's what I'm doing here. The simplifications are not assertions about the outside world: they are ways to restrict the domain of discourse, to keep it manageable. A more complicated analysis could probably handle the exceptional letters like Y too, but that would complicate the story too much for my present purpose.

With that caveat, am I right? Is it true that every solution of the SHIP–DOCK puzzle includes a word (in the new, restricted sense) with exactly *two* vowels (in the new, restricted sense)?

One way to investigate this is to look for other solutions, such as

SHIP
CHIP
CHOP
COOP
COOT
ROOT
ROOK
ROCK
DOCK

Here we find two vowels in COOP, COOT, ROOT, and ROOK. But even if a lot of individual solutions have two vowels somewhere, that doesn't prove that they all have to. A proof is a logical argument that leaves no room for doubt.

After a certain amount of experiment and thought, the "theorem" that I am proposing here starts to seem obvious. The more you think about how vowels can change their positions, the more obvious it becomes that somewhere along the way there must be exactly two vowels. But a feeling that something is "obvious" does not constitute a proof, and there's some subtlety in the theorem because some four-letter words contain *three* vowels, for instance, OOZE.

Yes, but . . . on the way to a three-vowel word, we surely have to pass through a two-vowel word? I agree, but that's not a proof either, though it may help us find one. *Why* must we pass through a two-vowel word?

A good way to find a proof here is to pay more attention to details. Keep your eye on where the vowels go. Initially, there is one vowel in the third position. At the end, we want one vowel in the second position. But—a simple but crucial insight—a vowel cannot change position in one step, because that would involve changing two letters. Let's pin that particular thought down, logically, so that we can rely on it. Here's one way to prove it. At some stage, a consonant in the second position has to change to a vowel, leaving all the other letters unchanged; at some other stage, the vowel in the third position has to change to a consonant. Maybe other vowels and consonants wander in and out, too, but whatever else happens, we can now be certain that a vowel cannot change position in one step.

How does the number of vowels in the word change? Well, it can stay the same; it can increase by 1 (when a consonant changes to a vowel), or it can decrease by 1 (when a vowel changes to a consonant). There are no other possibilities. The number of vowels starts at 1 with SHIP and ends at 1 with DOCK, but it can't be 1 at every step, because then the unique vowel would have to stay in the same place, position three, and we know that it has to end up in position two.

Idea: think about the *earliest* step at which the number of vowels changes. The number of vowels must have been 1 at all times before that step. Therefore it changes from 1 to something else. The only possibilities are 0 and 2, because the number either increases or decreases by 1.

Could it be 0? No, because that means the word would have no vowels at all, and by definition no "word" in our restricted sense can be like that. Therefore the word contains two vowels; end of proof. We've barely started analyzing the problem, and a proof has popped out of its own accord. This often happens when you follow the line of least resistance. Mind you, things really start to get interesting when the line of least resistance leads precisely nowhere.

It's always a good idea to check a proof on examples, because that way you often spot logical mistakes. Let's count the vowels, then:

SHIP	1 vowel
SHOP	1 vowel
SHOT	1 vowel
SLOT	1 vowel

SOOT	2 vowels
LOOT	2 vowels
LOOK	2 vowels
LOCK	1 vowel
DOCK	1 vowel

The proof says to find the first word where this number is not 1, and that's the word SOOT, which has two vowels. So the proof checks out in this example. Moreover, the number of vowels does indeed change by at most 1 at each step. Those facts alone do not mean that the proof is correct, however; to be sure of its correctness you have to check the chain of logic and make sure that each link is unbroken. I'll leave you to convince yourself that this is the case.

Notice the difference here between intuition and proof. Intuition tells us that the single vowel in SHIP can't hop around to a different position unless a new vowel appears somewhere. But this intuition doesn't constitute a proof. The proof emerges only when we try to pin the intuition down: yes, the number of vowels changes, but *when*? What must the change look like?

Not only do we become certain that two vowels must appear, we understand why this is inevitable. And we get additional information free of charge.

If a letter can sometimes be a vowel and sometimes a consonant, then this particular proof breaks down. For instance, with three-letter words there is a sequence:

SPA
SPY
SAY
SAD

If we count Y as a vowel in SPY but as a consonant in SAY (which is defensible but also debatable), then each word has a single vowel, but the vowel position moves. I don't think this effect can cause trouble when changing SHIP into DOCK, but that depends on a much closer analysis of the actual words in the dictionary. The real world can be messy.

Word puzzles are fun (try changing ORDER into CHAOS). This particular puzzle also teaches us something about proofs and logic. And about the idealizations that are often involved when we use math to model the real world.

There are two big issues about proof. The one that mathematicians worry about is, what *is* a proof? The rest of the world has a different concern: why do we need them?

Let me take those questions in reverse order: one now, and the other in a later letter.

I've begun to observe that when people ask why something is necessary, it is usually because they feel uncomfortable doing it and are hoping to be let off the hook. A student who knows how to construct proofs never asks what they're for. In fact, a student who knows how to do long multiplication in his head while doing a handstand also never asks why that's worth doing. People who enjoy performing an activity hardly ever feel the need to question its worth; the enjoyment alone is enough. So the student who asks why we need proofs is probably having trouble understanding them, or constructing his own. He is hoping you will answer, "There's no need to worry about proofs. They're totally useless. In fact, I've taken them off the syllabus and they won't come up in the test."

Ah, in your dreams.

It's still a good question, and if I leave it at what I've just said, I'm ducking the issue just as blatantly as any proofophobic student.

Mathematicians need proofs to keep them honest. All technical areas of human activity need reality checks. It is not enough to believe that something works, that it is a good way to proceed, or even that it is true. We need to know *why* it's true. Otherwise, we don't know anything at all.

Engineers test their ideas by building them and seeing whether they hold up or fall apart. Increasingly they do this in simulation rather than by building a bridge and hoping it won't fall down, and in so doing they refer their problems back to physics and mathematics, which are the sources of the rules employed in their calculations and the algorithms that implement those rules. Even so, unexpected problems can turn up. The Millennium

Bridge, a footbridge across the Thames in London, looked fine in the computer models. When it opened and people started to use it, it suddenly started to sway alarmingly from side to side. It was still safe, it wasn't going to fall down, but crossing it wasn't an enjoyable experience. At that point it became clear that the simulations had modeled people as smoothly moving masses; they had ignored the vibrations induced by feet hitting the deck.

The military learned long ago that when soldiers cross a bridge, they should fall out of step. The synchronized impact of several hundred right feet can set up vibrations and do serious damage. I suspect this fact was known to the Romans. No one expected a similar kind of synchronization to arise when individuals walked across the bridge at random. But people on a bridge respond to the movement of the bridge, and they do so in a similar way and at roughly the same time. So when the bridge moved slightly—perhaps in response to a gust of wind—the people on it started to move in synchrony. The more closely the footsteps of the people became synchronized, the more the bridge moved, which in turn increased the synchronization of the footsteps. Soon the whole bridge was swaying from side to side.

Physicists use mathematics to study what they amusingly call the real world. It is real, in a sense, but much of physics addresses rather artificial aspects of reality, such as a lone electron or a solar system with only one planet.

Physicists are often scathing about proofs, partly out of fear, but also because experiment gives them a very effective way to check their assumptions and calculations. If an intuitively plausible idea leads to results that agree with experiment, there's not much point in holding the entire subject up for ten, fifty, or three hundred years until someone finds a rigorous proof. I agree entirely. For example, there are calculations in quantum field theory that have no rigorous logical justification, but agree with experiment to an accuracy of nine decimal places. It would be silly to suggest that this agreement is an accident, and that no physical principle is involved.

The mathematicians' take on this, though, goes further. Given such impressive agreement, it is equally silly not to try to find out the deep logic that justifies the calculation. Such understanding should advance physics directly, but if not, it will surely advance mathematics. And mathematics often impinges indirectly on physics, probably a different branch of physics, but if so, it's a bonus.

So that's why proofs are necessary, Meg. Even for people who would prefer not to be bothered with them.

9

■ Can't Computers
Solve Everything?

Dear Meg,

Yes, computers can calculate much faster than humans, and much more accurately, which I gather has led some of your friends to question the value of your studies. There is a point of view that this makes mathematicians obsolete. I can reassure you that we're not.

Anyone who thinks computers can supplant mathematicians understands neither computing nor mathematics. It's like thinking we don't need biologists now that we have microscopes. Possibly the underlying misconception is that math is just arithmetic, and since computers can do arithmetic faster and more accurately than humans, why do we need the humans? The answer, of course, is that math is not just arithmetic.

Microscopes made biology more interesting, not less, by opening up new ways to approach the subject. It's the same with computers and mathematics. One very interesting and important thing computers have done for the mathematician is to make it possible to carry out experiments quickly and rapidly. These experiments test conjectures, occasionally reveal that what we were hoping to prove is wrong, and—with increasing frequency—perform gigantic calculations that allow us to prove theorems that would otherwise be out of reach. Sometimes the people who think math consists of big sums are right.

Take, for example, the Goldbach conjecture. In 1742 Christian Goldbach, an amateur mathematician, wrote to Leonhard Euler that as far as he could verify, every even number is a sum of two primes. For instance, $8 = 3 + 5$, $10 = 5 + 5$, $100 = 3 + 97$. Calculating by hand, Goldbach could test his conjecture on just a small list of numbers. On a modern computer you can quickly test billions of numbers: the current record is 2×10^{17}. Every time anyone has carried out such a test, they have found that Goldbach was right. However, no proof of his conjecture (which would turn it into a theorem) has yet been found.

Why worry? If a billion experiments agree with Goldbach, surely what he said must be true?

The problem is that mathematicians use theorems to prove other theorems. A single false statement in principle ruins the whole of mathematics. (In practice, we would notice the falsehood, isolate the offending statement, and avoid using it.) For example, the number π is rather annoying, and it would be nice to get rid of it. We might decide that there's no real harm in replacing π with 3 (as some say the Bible does, but only if you take an obscure passage extremely literally) or by 22/7. And if all you want to use π for is to calculate perimeters of circles and the like, a good enough approximation will do no harm.

However, if you really think that π is *equal* to 3, the repercussions are much greater. A simple line of reasoning reveals an unintended consequence. If $\pi = 3$, then $\pi - 3 = 0$, and dividing both sides by $\pi - 3$ (which, thanks to Archimedes, we know is nonzero, so division is permitted), we get $1 = 0$. Multiplying by any number we wish, we prove that all numbers are zero; it follows that *any two numbers are the same*. So when you go into your bank to draw $100 from your account, the teller can give you $1 and insist that this makes no difference; and in fact it doesn't, because you can then walk into Neiman Marcus or a Rolls-Royce dealership and explain that your $1 equals a million. More interestingly, murderers should not be jailed, because killing one person is the same as killing none; on the other hand, someone who has never touched drugs in their life should be imprisoned, since possession of no cocaine is the same as possessing a million tons of it. And so on . . .

Mathematical facts fit together and lead, via logic, to new facts. A deduction is only as strong as its weakest link. To be safe, all weak links must be removed. It would therefore be dangerous to verify Goldbach's conjecture on a computer for numbers up to, say, twenty digits, and conclude that it must be *true*.

Surely, you think, Ian's being a bit pedantic. If a statement is true for such big numbers, it has to be true for any numbers, no?

No. It doesn't work like that. For a start, twenty digits, in the mathematical universe, is tiny. The great ocean of numbers stretches to infinity, and even a number with a billion digits is, in some contexts, small. A classic example occurs in prime number theory. Although there is no evident pattern to the sequence of primes, there are statistical regularities. Sometime before 1849, when he wrote a letter about the discovery, Carl Friedrich Gauss found a good approximate formula relating the number of primes less than some specified number to the "logarithmic integral" of that number. It was soon noticed that the approximation always seems to be slightly larger than the correct value. Again, computer experiments verified that property for billions of numbers.

But the generalization is false. In 1914, John Littlewood proved that the correct value and the logarithmic integral approximation swap places infinitely often. But no one knew of a *specific* number for which the approximation was smaller than the correct value, until Littlewood's student Samuel Skewes proved, in 1933, that such a number must have at most $10^{10000000000000000000000000000000000}$

digits. That's thirty-four zeros in the *exponent*. Moreover, his proof involved an assumption, a notorious unproved statement known as the Riemann hypothesis. In 1955 Skewes proved that if you do not assume the Riemann hypothesis, then those thirty-four zeros in the exponent must be increased to one thousand zeros. And this gigantic number, please recall, is not the one we seek: it is the *number of digits* that number possesses.

Skewes's number has since been reduced to 1.4×10^{316}, which is tiny by comparison.

With numbers this large, the kind of experiment we can perform on a computer is totally irrelevant. And in number theory, that size is rather typical.

It wouldn't matter if all we were trying to do was approximate primes in terms of logarithmic integrals. But mathematics deduces new facts from old ones. And as we saw with π, if the old fact is actually wrong, consequent deductions can destroy the entire basis of mathematics. No matter how extensive the evidence of computer calculations may be, we still have to use the grey matter between our ears. Computers can be valuable assistants, but only when a lot of human thought has gone into setting up the computations. We have not yet been made obsolete by our own creations.

10

■ Mathematical Storytelling

Dear Meg,

In my last letter I told you why proofs are necessary. Now I turn to the other question I raised: what is a proof?

The earliest recorded proofs, along with the notion that proofs are necessary, occur in Euclid. His *Elements*, written around 300 BC, laid out much of Greek geometry in a logical sequence. It begins with two types of fundamental assumptions, which Euclid calls axioms and common notions. Both are basically a list of assumptions that will be made. For instance, axiom 4 states that "all right angles are equal," and common notion 2 states that "if equals be added to equals, the wholes are equal." The main difference is that the axioms are about geometric things and the common notions are about equality. The modern approach lumps them all together as axioms.

These assumptions are stated in order to provide a logical starting point. No attempt is made to prove them; they are the "rules of the game" for Euclidean geometry. You are free to disagree with the assumptions or invent new ones, if you wish; but if you do, you are playing a different game with different rules. All Euclid is trying to do is make the rules of *his* game explicit, so that the players know where they stand.

This is the axiomatic method, which remains in use today. Later mathematicians observed gaps in Euclid's logic, unstated assumptions that should really be included as axioms. For example, any line passing through the center of a circle must meet its circumference if the line is extended sufficiently far. Some tried to prove Euclid's most complex axiom, that parallel lines neither converge nor diverge from the others. Eventually, Euclid's good taste was demonstrated when it was realized that all such attempts are bound to fail. To muddy the philosophical waters, over the centuries, various deep difficulties in the axiomatic approach have appeared, such as Gödel's discovery that if mathematics is logically consistent, then we can never prove it to be so. We can live with Gödelian uncertainties if we have to, and we *do* have to.

Textbooks of mathematical logic base their descriptions of "proof" on Euclid's model. A proof, they tell us, is a finite sequence of logical deductions that begins with either axioms or previously proved results and leads to a conclusion, known as a *theorem*. Provided each step obeys the rules of logical inference—which can be found in textbooks on elementary logic—then the theorem is proved.

If you dispute the axioms, you are also free to dispute the theorem. If you prefer alternative rules of inference, you are free to invent your own. The claim is only that with those rules of inference, acceptance of the stated axioms implies acceptance of the theorem. If you want to make π equal to 3, you have to accept that all numbers are equal. If you want different numbers to be different, you have to accept that π is not 3. What you can't do is pick and mix, having π equal to 3 but zero different from one. It's as simple and sensible as that.

This definition of "proof" is all very well, but it is rather like defining a symphony as "a sequence of notes of varying pitch and duration, beginning with the first note and ending with the last." Something is missing. Moreover, hardly anybody ever writes a proof the way the logic books describe. In 1999 I was musing on this discrepancy, having accepted an invitation to a conference in Abisko, Sweden, on "Stories of Science and the Science of Stories." Abisko is north of the Arctic Circle, in Lapland, and a group of about thirty science fiction writers, popular science writers, journalists, and historians of science were going to spend a week there seeking common ground. Wondering what I was going to say to them, I suddenly realized what a proof *really* is.

A proof is a story.

It is a story told by mathematicians to mathematicians, expressed in their common language. It has a beginning (the hypotheses) and an end (the conclusion), and it falls apart instantly if there are any logical gaps. Anything routine or obvious can safely be omitted, because the audience knows about such things and wants the narrator to get on with the main story line. If you were reading a spy novel and the hero was dangling above a chasm on the end of a burning rope hanging from a helicopter, you would not want to read ten pages on the force of gravity and the physiological effects of a high-velocity impact. You'd want to find out how he saves himself. It is the same with proofs. "Don't waste time solving the quadratic, I know how to do that. But tell me, *why* do its solutions determine the stability of the limit cycle?"

In my paper (reprinted in *Mission to Abisko*) I said this: "If a proof is a story, then a memorable proof must tell a ripping yarn. What does that tell us about how to construct proofs? Not that we need a formal language in which every tiny detail can be checked algorithmically, but that the story line should come out clearly and strongly. It isn't the syntax of the proof that needs improvement: it's the semantics." That is, the essence of a proof is not its "grammar" but its *meaning*.

In that paper, I contrasted this admittedly vague and woolly notion with a much more formal one, the idea of a "structured proof," advocated by the computer scientist Leslie Lamport. Structured proofs make explicit every step in the logic, be it deep or trivial, clever or obvious. Lamport makes a strong case in favor of structured proofs as a teaching aid, and there's no doubt that they can be very effective in making sure that students really do understand details. Part of that case is an anecdote: a famous result called the Schröder–Bernstein theorem. Georg Cantor had found a way to count how many members a set has, even when that set is infinite, using a generalized type of number that he called a "transfinite number." The Schröder–Bernstein theorem tells us that if two transfinite numbers are less than or equal to each other, then they must actually be equal. Lamport was teaching a course based on the classic text *General Topology* by John Kelley, which includes a proof of the theorem, but when the extra details needed for the students were added, Kelley's proof turned out to be wrong.

Years later, Lamport could no longer locate the error. The proof seemed obviously correct. But five minutes spent writing down a structured proof revealed the mistake again.

I was worried, because I'd put a proof of the Schröder–Bernstein theorem into one of my own texts. I looked up Kelley's proof for myself but could not discover a mistake. So I e-mailed Lamport, who suggested that I write down a structured proof. Instead, I worked my way very systematically through Kelley's argument, in effect creating a structured proof in an informal way, and eventually I spotted the error.

There is a classic proof of the Schröder–Bernstein theorem that starts with two sets, corresponding to the two transfinite cardinals. Each set is split into three pieces, using a notion of "ancestor" invented purely for this particular proof, and the pieces are matched. In effect, this proof tells a story about the two sets and their component pieces. It's not the most gripping of stories, but it has a clear plot and a memorable punch line. Fortunately, I had used the classic proof in my textbook, and not Kelley's reworking of it. Because Kelley had told the wrong story. Attempting, I suspect, to simplify the classic version, he had overdone things and violated Einstein's dictum: "as simple as possible, but not more so."

The presence of this mistake supports Lamport's view about the value of structured proofs. But, to quote my paper, "There's another interpretation, not contradictory but complementary, which is that *Kelley told a good story badly*. It's rather as if he'd introduced the Three Musketeers as Pooh, Piglet, and Eeyore. Some parts of the story would [still] have made sense—their inseparable companionship, for instance—but others, such as the incessant swordplay, would not. . . ."

If we think of a proof in the textbook sense, all proofs are on the same footing, just as all pieces of music look like tadpoles sitting on a wire fence when expressed in musical notation, unless you are an expert and can "hear" sheet music in your head. But when we think of a proof as a story, there are good stories and bad ones, gripping tales and boring ones, like stirring or insipid music. There is an aesthetic of proof, so that a really good story can be a thing of beauty.

Paul Erdős had an unorthodox line on the beauty of proofs. Erdős was an eccentric and brilliant mathematician who collaborated with more people than anyone else on the planet; you can read his life story in Paul Hoffman's *The Man Who Loved Only Numbers*. Anybody who coauthored a paper with him is said to have an "Erdős number" of one. Their collaborators have Erdős number two, and so on. It's the mathematician's version of the Oracle of Kevin Bacon, in which actors are linked to Bacon by their appearances in the same movies, or by their appearances with actors who've appeared with Bacon, and so on. My Erdős number is three. I never collaborated with Erdős, and I'm not on the list of people with Erdős number two, but one of my collaborators is.

Anyway, Erdős reckoned that in Heaven, God had a book that contained all the best proofs. If Erdős was really impressed by a proof, he declared it to be from "The Book." In his view, the mathematician's job was to sneak a look over God's shoulder and pass on the beauty of His creation to the rest of His creatures.

Erdős's deity's Book is a book of stories. I ended my Abisko talk like this: "Psychologists now tell us that without emotional underpinnings, the rational part of our mind doesn't work. It seems that we can only be rational about things if we have an emotional commitment to such a recently evolved technique as rationality . . . I don't think I could get very emotional about a structured proof, however elegant. But when I can really feel the power of a mathematical story line, something happens in my mind that I can never forget . . . I'd rather we improved the storytelling of proofs, instead of dissecting them into bits that can be placed in stacks of file cards and sorted into order."

11

■ Going for the Jugular

Dear Meg,

If you want to make a real name for yourself as a mountaineer, you have to conquer a peak that no one else has climbed. If you want to make a real name for yourself as a mathematician, there is no better way than to vanquish one of the subject's long-standing unsolved problems. The Poincaré conjecture, the Riemann hypothesis, the Goldbach conjecture, the twin primes conjecture.

I don't advise being that ambitious when you are working for a PhD! Big problems, like big mountains, are dangerous. You could spend three or four years doing extremely clever things, but fail to achieve your goal and end up with nothing. Math differs from the other sciences in this regard: if you carry out a series of chemistry experiments, you can always write up the results whether or not they confirm your thesis adviser's theories. But in math, you normally can't write a thesis saying, "Here's how I tried to solve the problem, and here's why it didn't work."

Even professionals have to treat big problems with considerable respect. Universities nowadays expect their faculty to be productive, and they tend to measure productivity in terms of publications per year. If you publish nothing for five years and then solve the Poincaré conjecture, you'll be set for life, assuming you are allowed to keep your job while you are doing it. If you publish nothing for five years and then fail to solve the Poincaré conjecture, you'll be out on your ear.

The sensible compromise is to spend some of your time working on a big problem and the rest working smaller, solvable, but still worthwhile problems. It would be wonderful to live in a world where it was possible to focus solely on the big issues, but we don't. Nevertheless, a few brave souls have managed to find a way to do just that, and succeed. In their hands, the time-honored conjecture acquires a proof, and becomes a theorem.

In my last letter I told you that a proof is a story. It is often said that there are basically only seven plots for a novel, and the ancient Greeks knew them all. There seem to be relatively few narrative lines for mathematical proofs, too, but the ancient Greeks knew only one: the short, sweet, compellingly clever argument of Euclid's that made QED a household word.

What, then, are we to make of mathematical proofs that are hundreds or even thousands of pages long? Or proofs that involve months of calculation on a big network of computers? More and more of these daunting beasts are coming into existence, usually as solutions to one of those big, open problems. Instead of the short, compelling story line known to the Greeks, these proofs are epics, or worse, long cycles of stories whose main thread may be submerged in intricate subplots for chapters at a time. What happened to Erdős's vision of the beauty of God's mathematical creation? Are these mammoth proofs really necessary? Are they so vast only because mathematicians are too stupid to find the short, elegant versions written in The Book?

Wiles's landmark proof of Fermat's last theorem amounted to about one hundred pages of highly technical mathematics, prompting the science journalist John Horgan to write a provocative article titled "The Death of Proof." Horgan assembled a variety of reasons why proofs were becoming obsolete, including the rise of the computer, the disappearance of proofs from school math, and the existence of blockbusters like Wiles's. It was an interesting attempt to wrench defeat from the jaws of victory, to treat a historic achievement as bad news. Yes, we put a man on the moon, but look at all the valuable rocket fuel we had to use up.

Wiles's proof may be a blockbuster, but it tells a ripping yarn. He had to use massive mathematical machinery for so simple a question, much as a physicist needs a particle accelerator many miles in circumference to study a quark. But far from being sloppy and unwieldy, his proof is rich and beautiful. Those hundred pages have a plot, a story line. An expert can skim through the details and follow the narrative, with its twists and turns of logic, and its strong element of suspense: will the hero overcome the last theorem in the final pages, or will the ghost of Fermat continue to taunt the mathematical profession? No one declared literature dead because *War and Peace* was rather long or because *Finnegans Wake* was not being read in schools. Professional mathematicians can handle a hundred pages of proof. Even ten thousand pages—the total length of the classification theorem for finite simple groups, combining the work of dozens of people over a decade or more—does not daunt them.

There is no reason to expect every short, simple, and true statement to have a short, simple proof. In fact, there is good reason to expect the opposite. Gödel also proved that in principle, some short statements sometimes require long proofs. But we can never know, in advance, *which* short statements these are.

Pierre de Fermat was born in 1601. His father sold leather; his mother was the daughter of a family of parliamentary lawyers. In 1648 he became a king's councilor in the local parliament of Toulouse, where he served for the rest of his life, dying in 1665, just two days after concluding a legal case. He never held an academic position of any kind, but mathematics was his passion. The mathematical historian Bell called him "the Prince of amateurs," but most of today's professionals would be happy with half his ability. Though he worked in many fields of mathematics, Fermat's most influential ideas were in number theory, a subject that grew out of the work of Diophantus of Alexandria, who flourished around AD 250 and wrote a book called *Arithmetica*. It was about what are now called Diophantine equations, equations that must be solved in whole numbers.

One problem to which Diophantus gave a completely general answer is that of finding "Pythagorean triangles": two perfect squares whose sum is also a perfect square. Pythagoras's theorem tells us that such triples are the lengths of sides of a right triangle. Examples are $3^2 + 4^2 = 5^2$ and $5^2 + 12^2 = 13^2$. Fermat owned a copy of *Arithmetica*, which inspired many of his investigations, and he used to write down his conclusions in the margin. Sometime around 1637 he must have been thinking about the Pythagorean equation, and he asked himself what happens if instead of squares you try cubes or higher powers, for instance, $x^4 + y^4 = z^4$. But he couldn't find any examples. In the margin of his copy of *Arithmetica* he made the most famous note in the history of mathematics: "To resolve a cube into the sum of two cubes, a fourth power into two fourth powers, or in general any power higher than the second into two of the same kind, is impossible; of which fact I have found a remarkable proof. The margin is too small to contain it."

This statement has come to be known as his "last theorem," because for many years it was the only assertion of his that his successors had neither proved nor disproved. Nobody could reconstruct Fermat's "remarkable proof," and it seemed increasingly doubtful that he had really found one. But if he had possessed a proof, even though it couldn't fit in a margin, it would surely be concise and elegant enough to earn a place in God's Book? No one was writing blockbuster proofs in the seventeenth century. Yet for three and a half centuries, mathematician after mathematician failed to find Fermat's missing proof. Then, in the late 1980s, Wiles began an extended attack on the problem. He worked alone in the attic of his house, telling only a few select colleagues who were sworn to secrecy.

Wiles's strategy, like that of many mathematicians before him, was to assume that a solution existed and then play with the numbers algebraically in the hope that this would lead to a contradiction. His starting point was an idea emanating from the German mathematician Gerhard Frey, who realized that you could construct a type of cubic equation known as an elliptic curve from the three numbers occurring in the purported solution of Fermat's "impossible" equation. This was a brilliant idea, because mathematicians had been playing around with elliptic curves for more than a century and had developed plenty of ways of manipulating them. What's more, mathematicians then realized that the elliptic curve made from Fermat's roots would have such strange properties that it would contradict another conjecture—the so-called Taniyama–Shimura–Weil conjecture—that governs the behavior of such curves.

No one had ever proved the Taniyama–Shimura–Weil conjecture, though most mathematicians thought it was probably right. If it were right, of course, the roots of Fermat's equation would lead to a contradiction, showing that they could not exist. So Wiles took a deep breath and set about trying to prove the Taniyama–Shimura–Weil conjecture. For seven years, he brought every big gun of number theory to bear on it, until he eventually came up with a strategy that cracked it wide open. Although he worked alone, he didn't invent the whole area by himself. He kept in close touch with all new developments on elliptic curves, and without a strong community of number theorists creating a steady stream of new techniques, he probably would not have succeeded. Even so, his contribution is massive, and it is propelling the subject into exciting new territory.

Wiles's proof has now been published in full, and in print it comes to a bit over one hundred pages. Certainly too long to fit into a margin. Was it worth it?

Absolutely.

The machinery that Wiles developed to crack Fermat's last theorem is opening up entire new areas of number theory. Agreed, the story he had to tell was lengthy, and only experts in the area could hope to understand it in any detail, but it makes no more sense to complain about that than it would to complain that in order to read Tolstoy in the original, you have to be able to understand Russian.

1 2

▪ Blockbusters

Dear Meg,

No, I was not joking when I said that the classification of the finite simple groups runs to ten thousand pages. It is currently being simplified and reorganized, though. With luck and a following wind, it might shrink to only two thousand. Most of the proof was carried out by hand, and the ideas behind it were entirely the product of the human mind. But some key parts required computer assistance.

This is a growing trend, and it has led to a new kind of narrative style for proofs, the computer-assisted proof, which has come into existence only in the last thirty years or so. This is like the fast-food outlet that serves billions of dull, repetitive burgers: It does the job, but not prettily. There are often some clever ideas, but their role is to reduce the problem to a massive—but in principle routine—calculation. This is then entrusted to a computer, and if the computer says "yes," the proof is complete.

An example of this kind of proof turned up recently in relation to the Kepler problem. In 1611 Johannes Kepler was considering how spheres can be packed together. He came to the conclusion that the most efficient method—the one that packs as many balls as possible into a given region—is the one that greengrocers often use to stack oranges. Make a flat layer in a honeycomb pattern, then stack another such layer on top, with the oranges sitting in the depressions of the first layer, and continue like this forever. This pattern shows up in lots of crystals, and physicists call it the face-centered cubic lattice.

It is often said that Kepler's statement is "obvious," but anyone who thinks so does not appreciate the subtleties. For example, it is not even clear that the most efficient arrangement includes a flat plane of spheres. Greengrocers start stacking on a flat surface, but you don't *have* to. Even the two-dimensional version of the problem—showing that a honeycomb pattern is the most efficient way to pack equal circles in the plane—wasn't proved until 1947, when László Fejes Tóth finally did it. His proof is too complicated to belong in The Book, but it's all we have.

In 1998, Thomas Hales announced a computer-assisted proof of the Kepler conjecture that involved hundreds of pages of mathematics plus three gigabytes of supporting computer calculations. It has since been published in the *Annals of Mathematics*, the world's premier math journal, with one important reservation: the referees state that they have not been able to check every single step in the computations.

Hales's approach was to write down a list of all the possible ways to arrange suitable small clusters of spheres; then prove that whenever the cluster is not what you find in the face-centered cubic lattice, it can be "compressed" by rearranging the spheres. Conclusion: the only incompressible arrangement—the one that fills space most efficiently—is the conjectured one. This is how Tóth handled the two-dimensional case, and he needed to list about fifty possibilities. Hales had to deal with thousands, in three dimensions, and the computer had to verify an enormous list of inequalities—those three gigabytes of memory are what's needed to tabulate them all.

One of the earliest proofs to use this brute-force computer method was the proof of the four color theorem. About a century ago Francis Guthrie asked whether every possible two-dimensional map containing any arrangement of countries can be colored using only four colors, so that neighboring countries always get different colors. It sounds simple, but the proof was highly elusive. Eventually, in 1976, Kenneth Appel and Wolfgang Haken proved the four color theorem. By trial and error and hand calculations, they first came up with a list of nearly two thousand configurations of "countries" and invoked the computer to prove that the list is "unavoidable," meaning that every possible map must contain countries arranged in the same way as at least one configuration in the list.

The next step was to show that each of these configurations is "reducible." That is, each configuration can be shrunk down until a part of it disappears, leaving a simpler map. Crucially, the shrinking must ensure that if the simpler map left behind can be colored with four colors, the original one can be as well. Now remember that every possible map must contain at least one of the two thousand configurations. So even the simpler map that you have just created by this process must have another configuration that can be shrunk again and so on. The upshot is that if you can find a way to shrink every possible configuration, you have your proof. Matching every configuration with a way to "shrink" it like this involved a huge but routine computer calculation, which in those days took about two thousand hours on the fastest available computer. (Nowadays it takes maybe an hour.) But in the end, Appel and Haken had their answer.

Computer-assisted proofs raise issues of taste, creativity, technique, and philosophy. Some philosophers feel that with their brute-force methods, they are not proofs in the traditional sense. Yet this kind of massive but routine exercise is what computers were invented to do. It's what they're good at, and it's what humans are very poor at. If a computer and a human being both carry out a huge calculation and get different answers, the smart money is on the computer. But it must be said that any one bit of the proof, any one calculation by the computer, is usually trivial and extremely dull. It's only when you string them together that they're worth anything. If Wiles's proof of Fermat's last theorem is rich in ideas and form—like *War and Peace*—the computer proofs are more like telephone directories. In fact, for the Appel–Haken proof, and even more so Hales's proof, life is—literally—too short to *read* the whole thing in full detail, let alone check it.

Still, these proofs are not devoid of elegance and insight. You have to be pretty clever about how you set up the problem for the computer to tackle. What's more, once you know the conjecture is right, you can set about trying to find a more elegant way to prove it. This might sound strange, but it's well known among mathematicians that it's much easier to prove something you already know is true. In mathematical common rooms worldwide, you will occasionally hear someone suggest—only partly as a joke—that it might be a good idea to spread rumors that some important problem has been solved, in the hope that this might speed up its actual solution. It's a bit like crossing the Atlantic. For Christopher Columbus this was desperately hard, but it was easier for John Cabot, sailing just five years later, because he knew what Columbus had found.

Does this mean that eventually mathematicians may find God's proofs for Kepler and the four color theorem?

Maybe, but maybe not. It's a bit naive to imagine that every theorem that is simple to state must therefore have a simple proof. We all know that many tremendously difficult problems are deceptively easy to state: "land on the moon," "cure cancer." Why should math be different?

Experts often get rather passionate about proofs, either that the best-known one can't be simplified, or that alternative methods that someone is proposing can't possibly work. Often they're right, but sometimes their judgment can be affected by knowing too much. If you're an experienced mountain climber proposing to scale a high peak, with glaciers and crevasses and the like, the "obvious" path may be exceedingly long and complicated.

It's natural, too, to assume that the sheer cliff face, which seems to be the only alternative route, is simply unclimbable. But it may be possible to invent a helicopter that can swing you quickly and easily up to the top. The experts can see the crevasses and the cliff, but they may miss a good idea for the design of a helicopter. Occasionally someone invents such a piece of machinery out of the blue, and proves all the experts wrong.

On the other hand, think of Gödel. We know that some proofs simply *have* to be long, and perhaps the four color theorem and Fermat's last theorem are examples. For the four color theorem, it is possible to do some back-of-the-envelope calculations that show that if you want to use the current approach—finding a list of "unavoidable" configurations and then eliminating them one at a time by some "shrinking" process—then nothing radically shorter is possible. But that, in effect, is just counting the likely crevasses. It doesn't rule out a helicopter. So if these massive tomes are the best we can do, why did Fermat write what he did? Surely he can't have stumbled across a hundred-page proof, including within it a proof of a conjecture about elliptic curves that hadn't even been proposed yet, and scribbled down that it didn't quite fit into the margin.

A leading algebraic number theorist, Sir Peter Swinnerton-Dyer, has offered a simpler explanation of Fermat's claim: "I am sure that Fermat believed he had proved it; and indeed one can with fair confidence reconstruct his argument, including one vital but illegitimate step." It would be lovely to imagine that the great Fermat really was in possession of a proof, because the methods available to him would have been simpler than those used by Wiles. But it seems more likely that Fermat made a subtle but fatal error, one that would easily have gone undetected at that time.

13

■ Impossible Problems

Dear Meg,

Please don't try to trisect the angle. I'll send you some interesting problems to work on if you want to flex your research muscles already; just stay clear of angle trisection. Why? Because you'll be wasting your time. Methods that go beyond the traditional unmarked ruler and compass are well known; methods that do not cannot possibly be correct. We know that because mathematics enjoys a privilege that is denied to most other walks of life. In mathematics, we can prove that something is impossible.

In most walks of life, "impossible" may mean anything from "I can't be bothered" to "No one knows how to do it" to "Those in charge will never agree." The science fiction writer Arthur C. Clarke famously wrote that "When an elderly and distinguished scientist declares something to be possible, he is very likely right. When he declares it to be impossible, he is almost certainly wrong." (Clarke was writing in 1963, when most scientists, especially elderly and distinguished ones, were almost certainly "he.") But applied to mathematicians and mathematical theorems, Clarke's statement is plain wrong. A mathematical proof of impossibility is a virtually unbreakable guarantee.

I say "virtually" because sometimes the impossible can become possible if the question is subtly changed. Then, of course, it's not the same question. Archimedes knew how to trisect an angle using a *marked* ruler and compass.

My favorite simple impossible problem is a puzzle. Though it looks frivolous at first glance, it provides a lot of insight into logical inference in mathematics, and in particular into how we know that some tasks are impossible. The puzzle is this: given a chessboard with two diagonally opposite corners missing, can you cover it with thirty-one dominoes, each just the right size to cover two adjacent squares?

It must be understood that no "cheating" is allowed. The dominoes must not overlap, or be cut up, or anything like that.

The first question to ask is reasonably straightforward and comes naturally to any mathematician: is the area an obstacle to success? The total area of the mutilated chessboard is $64 - 2 = 62$ squares. The total area of the dominoes is $2 \times 31 = 62$ squares. So we have exactly the right number of dominoes to cover the chessboard. If we had been given only thirty, then calculating the total area would immediately have proved that the task is impossible. But we've been given thirty-one, so the area is not an obstacle.

Meg, I know you've done a lot of math, but it's just possible that you've never come across this puzzle. Puzzles do not feature prominently in university textbooks. Please try it. For the moment, don't think about it; just cut out some cardboard dominoes and try to fit them together.

Done that? Did you get anywhere?

No. You tried and tried but nothing worked. You can see why if you count the black and white squares.

Every domino, no matter where it is placed, covers one black and one white square of the chessboard. So any arrangement of nonoverlapping dominoes must cover the same number of black squares as white. But the mutilated chessboard has thirty-two black squares and thirty white ones. No matter how you arrange the dominoes, two black squares (at least) must always stay empty.

If instead, two adjacent corners are removed—one black, one white—this argument fails. And in fact, the puzzle can then be completed. But the "parity" argument, counting the numbers of squares of the two colors and comparing them, rules out the version that I posed to begin with. The task is impossible . . . period.

The deeper message behind this puzzle applies throughout mathematics. When a problem leads you to consider some huge number of possibilities—such as all the different ways the dominoes might be arranged—there is generally no practical way to deal with them one at a time. You must search for some common feature that does not change when you change the arrangement: an *invariant*.

Here, the first invariant we tried was area. If you rearrange the dominoes, their total area remains the same. But that invariant doesn't help here. So I resorted to a different invariant: the difference between the number of black squares and the number of white ones. This is always zero in any domino arrangement. So *no* arrangement in which the invariant is nonzero can be obtained by arranging dominoes according to our rules.

The proof leaves open the possibility that some arrangements with zero invariant might not be possible for other reasons. In fact, these do exist; maybe you can find some. The "area" invariant solves some puzzles but not others. So does the "parity" invariant, odd or even. The same is true of most invariants.

Now we must move on, from puzzles to serious mathematical problems. Remarkably, and delightfully, similar ideas still apply.

Trisecting the angle is a case in point. We now know—we have known it since Gauss's student Pierre Wantzel wrote down a proof in 1837—that it is impossible to trisect the angle using an unmarked straightedge and compass. That is, there is no geometric construction, employing the traditional implements in the traditional manner, that will divide any given angle into three exactly equal parts.

There are zillions of approximate constructions. None will be exact. I can say this without the slightest fear of contradiction and without examining any proposed method. We *know* it must contain a mistake. We don't know where or what the mistake is—and it may be very difficult to find—but we can be certain it's there.

This, I know, sounds arrogant. It can be very annoying to any would-be trisector. "How can they possibly know this when they haven't even looked at my proof?"

They know because such a construction has been proved to be impossible. If someone claimed they could run a mile in ten seconds, you wouldn't have to watch them try it to know that there has to be a trick. Maybe they "run" using rocket assistance. Maybe their "mile" is not measured in the orthodox manner and is no longer than a bus. Maybe something is funny about their clock. We don't need to know which to smell a rat.

It's like that in mathematics, but with a higher degree of certainty.

Very well: how do we know that angle trisection is impossible?

Although the problem concerned belongs to geometry, its resolution belongs to algebra. This is a standard ploy in mathematical research: try transforming your problem into something that is logically equivalent but lies in a different area of mathematics. If you're fortunate, the new area will permit the use of new techniques, casting new light on the problem. The idea of replacing geometry with algebra—or the reverse—goes back at least to René Descartes. In an appendix to his *Discours de la Méthode* of 1637, with the title *La Géometrie*, he outlined the use of coordinates to turn geometric forms into algebraic equations, and back again. Today we call them Cartesian coordinates in his honor.

The idea will be familiar to you, Meg. Any point in the plane can be characterized by two numbers, distances measured in two directions at right angles to each other. Horizontal and vertical, or north/south and east/west. A line, circle, or other curve is just a collection of points, a set of pairs of numbers. Any statement about those lines and curves can be converted to a corresponding statement about numbers, and such statements belong to the realm of algebra. Thus the fact that a circle has a radius of one unit, when converted into algebra using Pythagoras's theorem, becomes the fact that for any point on its circumference, the square of its horizontal coordinate, added to the square of its vertical coordinate, equals 1. In symbols, $x^2 + y^2 = 1$. This is the equation corresponding to that circle.

Every circle, every straight line, and every curve has a corresponding equation. And the points where, say, a circle meets a line are those pairs of numbers that satisfy both the equation for the circle and the equation for the line. Instead of drawing lines and curves and finding their intersections points, we can just solve equations. More importantly, instead of *thinking* about drawing lines and curves and finding their intersections points, we can think about solving the corresponding equations. And this is how we can prove that angles cannot be trisected in the manner specified.

Here's how it goes, stripped of technical details. Any geometrical construction begins with a collection of points and then constructs new points by one of three methods. Either we draw two lines through existing points and find out where those lines meet; or we draw one such line and find where it meets a circle centered on a known point and passing through another known point; or we draw two such circles and see where *they* meet. This small set of moves is a product of our tools: a straightedge makes only straight lines, and a compass makes only circles. Thus we build new points out of old ones, and we continue for a finite number of such moves and then stop.

This is another standard proof technique: break the problem down into the simplest possible parts.

It may appear that an angle does not fit this description. But an angle is specified by two lines that meet at a common point, and those two lines can be specified by the common point, another point on the first line, and another point on the second line. Three points suffice to define an angle. It takes only one further point to specify an angle one-third the size. But locating that fourth point may in principle require constructing a lot of auxiliary points along the way, and the claim is that none of these will actually help.

To see why, we apply another standard proof technique: examine each of the simplest steps and try to find its essential features.

Geometrically, there are three distinct steps: two lines, line plus circle, two circles. But if we convert those steps into algebra, we find that the first one is equivalent to solving a linear equation and the other two are equivalent to solving a quadratic equation. In a linear equation we are told that some multiple of the unknown, plus some number, is zero. In a quadratic equation we are told that some multiple of the square of the unknown, plus some multiple of the unknown, plus some number, is zero.

Linear equations are "special cases" of quadratic ones: the relevant multiple of the square of the unknown is the multiple by zero. So all three steps boil down to solving quadratic equations.

Methods for solving quadratic equations were known to the Babylonians in 2000 BC, and the basic idea is that you can always do it using square roots. In short, we have replaced "constructible using unmarked straightedge and compass" by "expressible by a sequence of square roots (and other arithmetical operations such as addition and subtraction)." That characterizes all possible points that can arise from geometric constructions.

Suppose that some angle can be trisected using such a construction. Then the coordinates of the corresponding point—the one associated with an angle one-third the size—must be expressible by a sequence of square roots. Is that possible? Well, we know something about that new point, namely, that its coordinates are given by a cubic equation, one that also involves the cube of the unknown. This observation comes from trigonometry, where there is a standard formula relating the sine of an angle to the sine of three times that angle.

The whole shebang, then, reduces to a simple question: given a number that you know is the solution of a cubic equation, is it possible to express that number using only square roots? The intuition is that there is a mismatch here: no sequence of steps involving the number 2 should give rise to the number 3. A close examination of the properties of solutions of equations leads to an invariant, known as the *degree*. This has nothing to do with "degree" as a way to measure angles; it is a whole number specifying the type of equation that is being solved. Simple properties of the degree prove that the only time that you can solve a cubic equation using nothing more elaborate than square roots is when the cubic equation breaks down into a linear equation and a quadratic, or three linear ones.

However, a short calculation shows that with rare exceptions, the cubic equation associated with angle trisection is not like that. It does not break down. In particular, this is the case when the initial angle is 60°, for example. Therefore the cubic equation cannot be solved *exactly* using only square roots. Indeed, if it could be solved in that manner, then the integer 3 would have to be an even number. But of course it isn't.

I've left out the details—you can find them in many standard algebra texts if you want to see them—but I hope the story is clear. By transforming the geometry into algebra, we can reformulate angle trisection (indeed any construction) as an algebraic question: can some number associated with the desired construction be expressed using square roots? If we know something useful about the number concerned—here, that it is given by a cubic equation—then we may be able to answer the algebraic version of the question. In this case, the algebra rules out any possibility that such a construction exists, thanks to the invariant known as the degree.

It's not a matter of being clever: however cunning you are, your purported construction will necessarily be inexact. It might be very accurate (sadly, most attempts aren't; have a look at Underwood Dudley's *A Budget of Trisections*), but it cannot be exact. It's not a matter of finding other methods for trisecting angles, either: those are already known, and that wasn't the question. I always tell anyone who sends me an attempted trisection that I'm not concerned about their being wrong, and neither should they be. The problem is that if they are *right*, then a direct consequence of their proof is the fact that 3 is an even number.

Do they really want to go into the history books as claiming that?

It doesn't put them off, mind you. No rational argument ever diverts a true trisector from their innate certainty that they are right.

The "degree" invariant also explains why a regular seventeen-sided polygon can be constructed but a seven-sided one cannot. The corresponding degrees turn out to be one fewer than the number of sides: sixteen and six. Because 16 is the fourth power of 2, the seventeen-gon can be constructed by solving four successive quadratics. But 6 is not a power of 2, so no construction exists in that case. In my experience, angle trisectors seldom object to this deduction, although ironically, it implies that a valid trisection of the angle would lead directly to a construction of the regular seven-gon.

There are many other impossible problems in mathematics. Angle trisection is one of three famous "problems of antiquity" credited to ancient Greek geometers, unfortunately without much historical justification, because the limitation to unmarked straightedge and compass was a later addition. The Greeks knew how to solve all three problems using more complex instruments. But it's true that this was the only way they knew how to solve them. Later mathematicians wondered whether anyone could do better, and eventually realized that they couldn't.

The other two problems of antiquity are squaring the circle and duplicating the cube. That is, using the traditional methods, construct a square whose area is equal to that of a given circle, or a cube with twice the volume of a given one. In modern terms, these problems ask for constructions for π and the cube root of two, respectively. They can be proved impossible by similar methods. In fact, the cube root of two evidently satisfies a cubic equation: its cube is two. And π satisfies no algebraic equation whatsoever; but that's another story.

14

■ The Career Ladder

Dear Meg,

Don't mention it. I am always happy to treat you to a meal whenever we find ourselves in the same city, which, if you're genuinely intent on pursuing a career in research, could be increasingly often.

But let me play devil's advocate for a moment. It's important to ask yourself whether you want to stay at university because that's where you feel most comfortable. You should not be looking for "comfortable" at your age.

Being a research mathematician is akin to being a writer or an artist: any glamour that's apparent to outsiders fades quickly in the face of the job's frustration, uncertainty, and hard, often solitary, work. You can't expect your occasional moments in the spotlight to make it all worthwhile. Unless you're more superficial than I believe you are, they can't possibly. Your satisfaction must come from the high you get when you suddenly, for the first time, understand the problem you're working on and see your way to a solution. I use the word "high" advisedly. You need to be something like an addict for this feeling to provide sufficient recompense for all the work.

Here's the paradox: though much of a mathematician's work is solitary, even lonely, the most important aspect of your research is not the field you choose or the problems you embark on but how you deal with the people around you.

When you set out to earn a PhD, you do not do so alone. Your fellow students will constitute an important support group; your department will function as your clan within the larger tribe of mathematicians around the world; above all, you will have a thesis adviser (or supervisor, as we say in the UK). Normally he or she will be an established expert with a solid track record in the area you plan to study. Sometimes it will be someone who completed their own PhD only a few years ago, in which case there will probably be a second, more senior adviser to add experience to the mix.

A young adviser is often an excellent choice. They are usually bubbling over with ideas, and having just come through the academic mill themselves, they will probably be more sympathetic to your struggles.

In the April 1991 issue of *The Psychologist*, my sociologist friend Helen Haste analyzed the patterns of gift-giving among the remote and backward peoples known as "academics." The article was an anthropological spoof, but it made some telling points. The gifts were copies of research papers, and the article classified academics into a six-rung career ladder, plus one unorthodox role that sticks out sideways.

You are about to join the first rung of the ladder by becoming a DXGS: Dr. X's graduate student. From there you will, I'm sure, progress rapidly to PYR, promising young researcher, and thence to ER, established researcher. If you elect to remain in academia, the succeeding grades are SS (senior scientist), GOP (grand old person), and EG (emeritus guru).

As a DXGS you will not yet have produced any ritual gifts of your own, and so cannot present them to anyone. You can request them, but normally only from your peers. When performing before the tribe—that is, giving seminars—you will repeatedly invoke two ancestors, a major theorist and your thesis adviser. The PYR is more relaxed and understands the rituals better. He or she will still invoke those two ancestors, but briefly and often as mere footnotes. Astute PYRs invoke SSs instead. They travel to tribal meetings (conferences) so laden with gifts that the journey is more like a pilgrimage, and dispense them liberally. They also feel able to request gifts from their seniors, though not too often and always politely. The ER seldom refers to a major theorist, preferring to mention only ancestors who are currently active, but—tellingly—an ER may also mention progeny, to prove that he or she has them. The ER does not bring gifts to the tribal meeting, having cleverly dispensed them in advance to the tribe's inner circle.

The SS invokes a major theorist frequently, with the goal of supplanting him or her by being seen to have made important advances over the major theorist's ideas. The SS never gives or receives in public but expects to receive many gifts by more covert means. The GOP sits at the pinnacle of the gift-giving hierarchy, offering no gifts but requesting them from everyone, especially juniors. The EG is invoked as an ancestor by almost everyone, but takes part in absolutely no exchanges of gifts.

The role that does not fit into this sequence—indeed, does not fit anywhere, which is its raison d'être, is the maverick guru (MG). Helen has this to say about MGs: "The Maverick Guru has an important symbolic role, having curious magical powers that cause fear and fascination in the community. The MG is outside the mainstream orthodoxy, but engaged in criticizing it. The MG cannot be invoked as an ancestor by junior members of the community who intend to stay within the mainstream ... An erstwhile MG rapidly becomes a GOP."

I mention all this because you need to appreciate your place within the tribe, and because your progression from DXGS to PYR to ER depends heavily on your choice of X, who should be either an ER, an SS, or perhaps a GOP. *Do not choose an MG*, no matter how attractive that option may seem, unless you intend your entire career to operate outside the conventional ladder. And on the whole, I advise you to stay clear of GOPs. Trust me: I wanted to become an MG but I think I've ended up as a GOP instead. A GOP will have an impressive record, but much of it will date to the dim and distant past: five years ago, or even longer. The older an academic becomes, the more intellectual baggage he carries. GOPs' minds tend to run along familiar grooves, and although they do this with impressive ease and confidence, their students may miss out on the really new ideas that are the lifeblood of research. Some GOPs, though, make excellent advisers despite that, usually those who are close to being MGs but aren't quite.

EGs never have students.

My adviser was an SS in the field of group theory—the formal mathematics of symmetry—named Brian Hartley. He was young, but not too young. I didn't choose him, and he didn't exactly choose me either; I chose the field, and the system allocated me to him because he was in that field. There were four or five alternative choices. Any of them would have worked—I later got to know them all well, as colleagues—but my research would have been very different. I was very lucky to get Brian, who put me onto a problem—more an entire program—that really suited my interests and abilities. He was brilliant. He saw me regularly, was always available if I got stuck, and he was hardly ever stumped for an idea.

Brian was, I think, slightly taken aback when I marched into his office on day one of my PhD course and demanded a research problem. Usually it takes students longer to get going. But within half an hour he had given me one—arising from one of his own papers, my first receipt of a gift—and it turned out to be a beauty. The program of research was to study a special type of group that a Russian mathematician, Anatoly Ivanovich Malcev, had associated with a different mathematical structure called a Lie algebra. This structure was first developed over a century ago by the Norwegian Sophus Lie, but (despite its name) it was mainly studied in the context of analysis, not abstract algebra. So Malcev's purely algebraic version was a new point of view. Like many Russians at the time, he had sketched the ideas but not developed them in detail. My problem was to take Malcev's thoughts and conjectures, and fill in the necessary proofs and other connections, in effect, to turn a set of sketches and renderings into a finished blueprint of a building. It took me three months, and I got hooked by Lie theory. I ended up writing my entire thesis on Lie algebras.

Brian's influence did not stop at research problems. He and his wife, Mary, entertained me and other PhD students at their home. Occasionally he invited me to join him at some jazz performance in a local pub. He was an academic father figure, a mentor, and a friend. In 1994 he died, unexpectedly, at the age of 55 while walking in the hills near Manchester. I wrote his obituary for *The Guardian*. It ended like this: "I last saw Brian a few weeks ago, at a meeting to mark the sixtieth birthday of a mutual friend. He had just taken up a much-prized fellowship that would relieve him of all teaching duties for five years to concentrate on research. He went out with his boots on, both literally, while walking in his beloved hills, and metaphorically. And that is how we shall all remember him."

I still find it hard to accept that he's gone.

As I say, I was lucky. The system assigned me the ideal adviser. But you can do better than that. Don't leave it up to chance: *choose* your thesis adviser. Read the literature, talk to people in the field, find out who has a good reputation and—crucially—who is good with students. Draw up a short list. Visit them; in effect, *interview* them. Then trust your instincts. And remember, you don't want a GOP who ignores you; you want a close personal relationship.

Dare I add, not *too* close? The cliché of faculty sleeping with their students exists because it does happen. Someone once observed that the more subjective the discipline, the better the faculty dressed, and a similar principle seems to apply to illicit sex. Mathematicians, by and large, indulge in rather less of it, perhaps because we dress so badly. In any case, everyone knows it's unprofessional, and nowadays there are sexual harassment laws. Enough said. For recreation and affection, restrict yourself to fellow students, or people from off campus, please.

A standard joke states that mathematical ability is typically passed from father to son-in-law (or, these days, from mother to daughter-in-law). The point was that a male PhD student often married his adviser's daughter. It's one way to meet people off campus. So your real ancestry can be affected by your mathematical ancestry.

Mathematicians are proud to trace their academic lineage through thesis advisers. Brian was my mathematical father, and Philip Hall my grandfather. Hall was of a generation for whom a PhD was unnecessary as a qualification for a university profession, but the most significant influence on his early work was William Burnside. Burnside can similarly be considered a mathematical son of Arthur Cayley, one of the most famous English mathematicians of Victorian times.

I remember these things and value them. I know where and how I fit into the family tree of mathematical thought. Arthur Cayley is as important an ancestor to me as any of my biological great-great-grandfathers.

Talent must be passed to succeeding generations. I've been thesis adviser to thirty students so far, twenty men and ten women. Since 1985, the proportions are fifty percent men, fifty percent women. I *know* women are just as good at math as men because I've watched both at close quarters. I am particularly proud of my mathematical daughters, most of whom hail from Portugal, where mathematics has long been viewed as a suitable activity for women. All of my Portuguese daughters have remained in mathematics. In fact, most of my graduate students have remained in mathematics, and every single one of them earned a PhD. However, one is now an accountant, several work in computing, and one owns an electronics company, or at least he did the last time I heard from him.

The rest of the world is now following Portugal's lead. In July 2005 the American Mathematical Society released the results of its 2004 *Annual Survey of the Mathematical Sciences*. Since the early 1990s, women have been receiving around 45 percent of all first degrees in math. Women received almost one-third of all U.S. doctorates in the mathematical sciences in the academic year 2003–2004, and one-quarter of those awarded in the top forty-eight math departments. In all, 333 women received math PhDs that year, the largest number ever recorded.

The idea that math is not a suitable subject for women is stone-cold dead. The career ladder is open to both sexes, though it is still unbalanced at the top end.

1 5

■ Pure or Applied?

Dear Meg,

When you're choosing a subject area as a first-year grad student in mathematics, many people will tell you that your biggest choice is whether to study pure or applied mathematics.

The short answer is that you should do both. A slightly longer version adds that the distinction is unhelpful and is rapidly becoming untenable. "Pure" and "applied" do represent two distinct approaches to our subject, but they are not in competition with each other. The physicist Eugene Wigner once commented on the "unreasonable effectiveness of mathematics" for providing insight into the natural world, and his choice of words makes it clear that he was talking about pure mathematics. Why should such abstract formulations, seemingly divorced from any connection to reality, be relevant to so many areas of science? Yet they are.

There are many styles of mathematics, and while these two styles happen to have names, they merely represent two points on a spectrum of mathematical thought. Pure mathematics merges into logic and philosophy, and applied math merges into mathematical physics and engineering. They are tendencies, not the extremes of the spectrum. By an accident of history, these two tendencies have created an administrative split in academic mathematics: many universities place pure mathematics and applied mathematics in different departments. They used to fight tooth and nail over every new appointment and committee representative, but lately they are getting on rather better.

As caricatured by applied mathematicians, pure math is abstract ivory-tower intellectual nonsense with no practical implications. Applied math, respond the pure-math diehards, is intellectually sloppy, lacks rigor, and substitutes number crunching for understanding. Like all good caricatures, both statements contain grains of truth, but you should not take them literally. Nevertheless, you will occasionally encounter these exaggerated attitudes, just as you will encounter throwbacks who still believe that women are no good at math and science. Ignore them; their time has passed. They just haven't noticed.

Timothy Poston, a mathematical colleague whom I have known for thirty-five years, wrote a penetrating article in 1981 in *Mathematics Tomorrow*. He observed—to paraphrase a complex argument—that the "purity" of pure mathematics is not that of an idle princess who refuses to sully her hands with good, honest work, but a purity of *method*. In pure math, you are not permitted to cut corners or leap to unjustified conclusions, however plausible. As Tim said, "Conceptual thinking is the salt of mathematics. If the salt has lost its savor, with what shall applications be salted?"

A middle ground, dubbed "applicable mathematics," emerged in the 1970s, but the name never really got established. My view is that all areas of mathematics are potentially applicable, although—as with equality in *Animal Farm*—some are more applicable than others. I prefer a single name, mathematics, and I believe it should be housed within a single university department. The emphasis nowadays is on developing the unity of overlapping areas of math and science, not imposing artificial boundaries.

It has taken us a while to reach this happy state of affairs.

Back in the days of Euler and Gauss, nobody distinguished between the internal structure of mathematics and the way it was used. Euler would write on the arrangement of masts in ships one day and on elliptic integrals the next. Gauss was immortalized by his work in number theory, including such gems as the law of quadratic reciprocity, but he also took time out to compute the orbit of Ceres, the first known asteroid. An empirical regularity in the spacing of the planets, the Titius–Bode law, predicted an unknown planet orbiting between Mars and Jupiter. In 1801 the Italian astronomer Giuseppe Piazzi discovered a celestial body in a suitable orbit and named it Ceres. The observations were so sparse that astronomers despaired of locating Ceres again when it reappeared from behind the sun. Gauss responded by improving the methods for calculating orbits, inventing such trifles as least-squares data fitting along the way. The work made him famous, and diverted him into celestial mechanics; even so, his greatest work is generally felt to have been in pure math.

Gauss went on to conduct geographical surveys and to invent the telegraph. No one could accuse him of being impractical. In applied mathematics, he was a genius. But in pure mathematics, he was a god.

By the time the nineteenth century merged into the twentieth, mathematics had become too big for any one person to encompass it all. People started to specialize. Researchers gravitated toward areas of mathematics whose methods appealed to them. Those who liked puzzling out strange patterns and relished the logical struggles required to find proofs specialized in the more abstract parts of the mathematical landscape. Practical types who wanted *answers* were drawn to areas that bordered on physics and engineering.

By 1960 this divergence had become a split. What pure mathematicians considered mainstream—analysis, topology, algebra—had wandered off into realms of abstraction that were severely uncongenial to those of a practical turn of mind. Meanwhile, applied mathematicians were sacrificing logical rigor to extract numbers from increasingly difficult equations. Getting *an* answer became more important than getting the right answer, and any argument that led to a reasonable solution was acceptable, even if no one could explain why it worked. Physics students were told not to take courses from mathematicians because it would destroy their minds.

Rather too many of the people involved in this debate failed to notice that there was no particular reason to restrict mathematical activity to one style of thought. There was no good reason to assume that either pure math was good and applied math was bad, or the other way around. But many people took these positions anyway. The pure mathematicians didn't help by being ostentatiously unconcerned about the utility of anything they did; many, like Hardy, were proud that their work had no practical value. In retrospect, there was one good reason for this, among several bad ones. The pursuit of generality led to a close examination of the structure of mathematics, and this in turn revealed some big gaps in our understanding of the subject's foundations. Assumptions that had seemed so obvious that no one realized they *were* assumptions turned out to be false.

For instance, everyone had assumed that any continuous curve must have a well-defined tangent, almost everywhere, though of course not at sharp corners, which is why "everywhere" was clearly too strong a statement. Equivalently, every continuous function must be differentiable almost everywhere.

Not so. Karl Weierstrass found a simple continuous function that is differentiable *nowhere*.

Does this matter? Similar difficulties plagued the area known as Fourier analysis for a century, to such an extent that no one was sure which theorems were right and which were wrong. None of that stopped engineers from making good use of Fourier analysis. But one consequence of the struggle to sort the whole area out was the creation of measure theory, which later provided the foundations of the theory of probability. Another was fractal geometry, one of the most promising ways to understand nature's irregularities. Problems of rigor seldom affect immediate, direct applications of mathematical concepts. But sorting these problems out usually reveals elegant new ideas, important in some other area of application, that might otherwise have been missed.

Leaving conceptual difficulties unresolved is a bit like using new credit cards to pay off the debts on old ones. You can keep going like that for some time, but eventually the bills come due.

The style of mathematical thinking needed to sort out Fourier analysis was unfamiliar even to pure mathematicians. All too often, it seemed the objective was not to prove new theorems but to devise fiendishly complicated examples that placed limits on existing ones. Many pure mathematicians were disturbed by these examples, deeming them "pathological" and "monstrous," and hoped that if they were ignored they would somehow go away. To his credit, David Hilbert, one of the leading mathematicians of the early 1900s, disagreed, referring to the newly emerging area as a "paradise." It took a while for most mathematicians to see his point. By the 1960s, however, they had taken it on board to such an extent that their minds were focused almost exclusively on sorting out the internal difficulties of the big mathematical theories. When your understanding of topology does not permit you to distinguish a reef knot from a granny knot, it seems pointless to worry about applications. Those must wait until we've sorted this stuff out; don't expect me to build a cocktail cabinet when I'm still trying to sharpen the saw.

It did look a bit Ivory Tower. But collectively, mathematicians had not forgotten that the most important creative force in mathematics is its link to the natural world. As the theories became more powerful and the gaps were filled, individuals picked up the new kit of tools and started using them. They began wading into territory that had belonged to the applied mathematicians, who objected to the interlopers and weren't comfortable with their methods.

Mark Kac, a probabilist with interests in many other areas of application, wrote an amusing and penetrating

analysis of the pure mathematicians' tendency to reformulate applied problems in abstract terms. He likened
their approach to the invention of "dehydrated elephants": technically difficult, but of no practical value.
My friend Tim Poston pointed out that this is a poor
analogy. It is actually rather easy to create a dehydrated
elephant. The important technical issue is quite different: it is to ensure that when you add water, you get back
a fully functioning elephant. Hannibal, he said, could
have done with a cartload of dehydrated elephants when
he marched on Rome.

Metaphors notwithstanding, Kac had a point: abstract reformulations are not an end in themselves. But
he ruined it completely by offering an example. My
wife's father made the same mistake in the 1950s when
he said, correctly, that most pop stars would never last,
but then offered as his example Elvis Presley. Kac's example of an archetypal dehydrated elephant was Steven
Smale's recasting of classical mechanics in terms of
"symplectic geometry." It would be too great a diversion
to explain this new kind of geometry, but suffice it to say
that Smale's idea is now seen as a profound application
of topology to physics.

Another vocal critic of abstraction in math was John
Hammersley. A severely practical man and a consum-
mate problem solver, Hammersley watched with dismay
as the "new math" of the 1960s took over school curric-
ula worldwide, and things like solving quadratic equa-

tions were dumped in favor of gluing Möbius strips together to see how many sides and edges they had. In 1968 he wrote a celebrated diatribe called "On the Enfeeblement of Mathematical Skills by 'Modern Mathematics' and by Similar Soft Intellectual Trash in Schools and Universities."

Like Kac, he had a point, but it would have been a much better point if he had not been so certain that anything he didn't like was trash. "Abstract" is a verb as well as an adjective; generalities are abstracted from specifics. It is best to teach the specifics before performing the abstraction. But in the late 1960s, educators were throwing out the specifics. They had convinced themselves that it was more important to know that $7 + 11 = 11 + 7$ than to know that either of them was 18, and even better to know that $a + b = b + a$ without having a clue what a and b were. I can understand why Hammersley was livid. But . . . oh dear. From today's vantage point he looks like a knee-jerk reactionary. It turns out that that "soft intellectual trash" consisted of useful and important ideas, but ideas best taught in a university, not high school. At its frontiers, mathematics had to become general and abstract: otherwise there could be no progress. Looking back on the 1960s from the twenty-first century, when the work of that period is bearing fruit, I think Hammersley failed to appreciate that new applications would need new tools, or that the theories being developed so assiduously by the pure mathematicians would be a major source of those tools.

What Hammersley denigrated forty years ago as "soft intellectual trash" is precisely what I use today to work on problems in fluid mechanics, evolutionary biology, and neuroscience. I use group theory, the fundamental language of symmetry, to understand the generalities of pattern formation and the application of those ideas to many areas of science. So do hundreds of others in math, physics, chemistry, astronomy, engineering, and biology.

People who are proud to be "practical" bother me just as much as those who are proud not to be. Both can suffer from blinkered vision. I am reminded of the chemist Thomas Midgley, Jr., who devoted much of his professional life to two major inventions: Freon and leaded gasoline. Freon is a chlorofluorocarbon (CFC), and this class of chemicals was responsible for the hole on the ozone layer and is now largely banned. Lead in gasoline is also banned, because of its adverse effects on health, especially that of children. Sometimes a narrow focus on immediate practicalities can cause trouble later. It is easy to be wise after the event, of course, and the catalytic reactions on ice crystals that made apparently stable CFCs have such a damaging effect on the upper atmosphere were difficult to anticipate. But leaded gasoline was always a bad idea.

It's fine for people to advocate their point of view on how math should be done. But they should not presume that there is only one good way to do math. I value diversity, Meg, and I urge you to do the same. I also value imagination, and I encourage you to develop yours and use it. It takes a strong mixture of imagination and skepticism to see that what's currently in fashion will not always be so, or that what your colleagues dismiss as a fad may be something considerably more. Today's trendy fashion sometimes turns out to be threaded with pure gold.

Keep your mind open, but not so open that your brains fall out.

Over the years, a number of new areas of math have emerged from a diversity of sources, inspired by questions in the real world, or extracted from abstract theories because someone thought they were interesting. Some have attracted media attention, including fractal geometry, nonlinear dynamics ("chaos theory"), and complex systems. Fractals are shapes that have detailed structure on all scales of magnification, like ferns and mountains. Chaos is highly irregular behavior (such as weather) caused by deterministic laws. Complex systems model the interactions of large numbers of relatively simple entities, such as traders in the stock market. In the professional literature and mathematical house magazines, you will occasionally find criticisms of these areas that have that all-too-familiar reactionary feel: dismissive of anything that hasn't had a century-long track record or that the critic does not work on. What has really annoyed the critics is not the content of these new areas but the media exposure, which their own area, so obviously superior, is not getting.

It's actually rather easy to assess the scientific influence of, for instance, fractals or chaos. All you have to do is read *Nature* or *Science* for a month, and you will see them being used to investigate such things as how molecules break up during a chemical reaction, how gas giant planets capture new moons, or how species in an ecosystem partition resources. The scientific community accepted them long ago, to the extent that their use is now routine and unremarkable. Yet some diehards, who apparently don't sample broader reaches of science, still dispute that these areas have any importance. I'm afraid that they are about twenty years out of date. You can't dismiss something as a nine-day wonder when it has survived for nine thousand days and is currently thriving.

These people need to get out more.

Both Kac and Hammersley were unusually creative in their own fields, where their attitudes were imaginative and forward-looking. So it is slightly unfair to hold them up as examples of reactionary mathematicians. They expressed attitudes that were common in their day. Kac made major advances in probability theory, and his "shape of a drum" paper is a gem. Hammersley's 2004 obituary in the *Independent on Friday* had this to say about his work: "Hammersley . . . posed and solved some beautiful problems, among the best of which are self-avoiding [random] walks and percolation. He was delighted to learn in retirement of the recognition accorded thereto by mathematicians and physicists, and of the enormous progress made since his own pioneering work." But it added, "Ironically, recent progress has been made via a general theory rather than by the type of hands-on technique favoured by Hammersley."

This may be ironic but it is also entirely predictable. Hammersley was from the make-do-and-mend generation of applied mathematicians. Nowadays, more attention is paid to having the right tools for the job.

We live in a world whose technological abilities, and needs, are exploding. New questions require new methods, and purity of method remains vital, however practical the context. So do intuitive leaps, when they lead in creative directions, even if at first there are no proofs: new mathematics paves the way to new understanding. Which brings me back to Wigner and his classic essay "The Unreasonable Effectiveness of Mathematics in the Natural Sciences." Wigner wasn't just wondering why mathematics is *effective* in informing us about nature. Many people have picked up on this aspect of the issue, and offered what I think is an excellent answer: whether or not any particular mathematician notices, the development of mathematics is, and has always been, a two-way trade between real-world problems and symbolic or geometric methods devised to obtain answers. *Of course* math is effective for understanding nature; that, ultimately, is where it comes from.

But I think Wigner was worried—or pleasantly surprised—by something deeper. There is no reason to be astonished if someone starts from a real-world problem—say, the elliptical orbit of Mars—and develops the mathematics to describe it. This is exactly what Isaac Newton did with his inverse square law of gravity, his laws of motion, and calculus. But it is much more difficult to explain why the same tools (differential equations in this instance) provide significant insights into unrelated questions of aerodynamics or population biology. It is here that the effectiveness of math becomes "unreasonable." It's like inventing a clock to tell the time and then discovering that it's really good for navigation, which actually happened, as Dava Sobel explained in *Longitude*.

How can an idea extracted from a particular real-world problem resolve some totally different problem?

Some scientists believe that it happens because the universe really is made from mathematics. John Barrow argues the case like this: "For the fundamental physicist, mathematics is something that is altogether more persuasive. The farther one goes from everyday experience and the local world, the correct apprehension of which is a prerequisite for our evolution and survival, the more impressively mathematics works. In the inner space of elementary particles or the outer space of astronomy, the predictions of mathematics are almost unreasonably accurate. . . . This has persuaded many physicists that the view that mathematics is simply a cultural creation is a woefully inadequate explanation of its existence and effectiveness in describing the world. . . . If the world is mathematical at its deepest level, then mathematics is the analogy that never breaks down."

It would be lovely if this were true. But there is a different explanation, less mystical, less fundamentalist. Possibly less convincing.

Both differential equations and clocks are *tools*, not answers. They work by embedding the original problem in a more general context, and deriving general methods to understand that context. This generality improves their chance of being useful elsewhere. This is why their effectiveness appears unreasonable.

You can't always know in advance what uses you'll find for a good tool. A round piece of wood mounted on an axle becomes a wheel, useful for moving heavy objects. Cut a groove in its circumference and drape a rope around it, and the wheel becomes a pulley with which you can not only move objects but lift them. Make the wheel out of metal instead of wood, add teeth instead of a groove, and you have a gear. Put your gears and pulleys together with a few other elements—a pendulum, some weights, a face abstracted from an ancient sundial—and you have a mechanism for telling time, which is something the wheel's original inventors could never have anticipated. The pure mathematicians of the 1960s were forging tools that could be used by everyone in the

1980s. I have great respect for the Hammersleys of this world, much as I respect a large Alsatian dog I meet in the street. My respect for the dog's teeth does not lead me to agree with its opinions. If everyone adopted the attitudes advocated by Kac and Hammersley, no one would develop the crazy ideas that create revolutions.

So: should you study pure math or applied math?

Neither. You should use the tools at hand, adapt and modify them to suit your own projects, and make new ones as the need arises.

16

Where Do You Get Those Crazy Ideas?

Dear Meg,

It's easy to make research sound glamorous: grappling with problems at the cutting edge of human thought, making discoveries that will last a thousand years . . . There is certainly nothing quite like it. All it requires is an original mind, time to think, a place where you can work, access to a good library, access to a good computer system, a photocopier, and a fast Internet connection. All of these will be provided for you as part of your PhD course, except for the first item, which you will have to provide for yourself.

This is, of course, the sine qua non, the thing without which all the other items are useless. Normally, students are not admitted to a PhD course unless they have shown some evidence of original thinking, maybe in a project or a master's thesis. Originality is one of those things that you either have or you don't: it can't be taught. It can be nurtured or suppressed, but there isn't an Originality 101 course that will anoint you as able to think new thoughts provided you have read the textbook and passed the exam.

In saying this, I recognize that I am at odds with the prevailing view among educational psychologists, which is that anybody can achieve anything provided they undergo sufficient training. Observing that talented musicians practice a lot, the psychologists have deduced that it is practice that causes talent, and generalized that assertion to all other areas of intellectual activity. But their beliefs are founded on bad experimental design. What they must do, to test their theory, is to start with lots of people who *lack* musical talent, say, the certifiably tone-deaf. Train half of them, keeping the other half as a control group, and show that the training produces lots of highly talented musicians while its absence (predictably) does not. I am sure that training can lead to some improvement. I do not believe it can produce a decent musician unless the talent was there to begin with.

I am no Mozart. I have some musical talent, but not enough, and it's not for lack of practice. Training can get me to a reasonable level of proficiency: as an undergraduate I played lead guitar in a rock band. But all the practice in the world could never turn me into a Jimi Hendrix or Eric Clapton, never mind Mozart. As Edward Bulwer-Lytton said, "Genius does what it must, and talent does what it can." I have just enough musical talent to know what's lacking.

I *do* have mathematical talent. Not at the Mozart level, but a big improvement on my guitar playing. By age ten I was already the best in my class at math, and believe me, it did not come from lots of regular practice. My dreadful secret is that I did very little work on math. I didn't have to. My classmates thought I must have put in hours and hours of effort in order to wipe the floor with them in the math tests, and I had enough sense not to set them straight. They would have killed me if they had known how little time I spent on the math homework, compared to their own strenuous efforts.

When I was an undergraduate, at Churchill College, Cambridge, I had a friend who was also taking a math degree. He worked twelve hours a day, every day. I went to lectures, scribbled notes, spent an hour or two a week working through the problem sheets, and that was it until it was time to revise the material for the end-of-year exams. In the British system at that time, there were no end-of-term tests. You waited until June and then you took exams on everything you had studied over the year. So I worked harder in April and May than I did the rest of the year. But while my friend was studying late into the night, I was down at the pub having a beer and playing darts. And what was his reward for all that training? He barely scraped a pass. Whereas I got first-class marks (the British equivalent of straight A's) and a College Scholarship.

It is true that talented people often train very hard. They have to, to stay at the pinnacle of their chosen field. A football player who did not spend hours every day on fitness training would quickly be replaced by one who did. But the talent has to be there initially in order for the training to be effective.

I suspect that psychologists overrate the role of training because they have fallen for a politically correct theory of child development that views all new young minds as "blank slates" upon which anything whatsoever can be written. This theory was comprehensively demolished by Steven Pinker in his book *The Blank Slate*, but definitive refutation has never been a match for fervent belief.

Anyway, Meg, since you have been accepted onto a PhD course, the mathematicians who run it clearly believe you possess sufficient originality to complete it successfully. I am in no doubt that you also possess another essential quality: commitment. You *want* to do research; you are hungry for it. One of my colleagues once said to me, "I really can't tell who the best mathematicians are, but I can tell who is *driven*." Some people believe that in career terms, once you assume a fairly ordinary level of competence, energy and drive actually matter more than talent.

Science fiction writers, another profession where originality is essential, are often asked, "Where do you get those crazy ideas?" The standard answer is, "We make them up." I've written sci-fi novels, and I concur.

But authors do not make up ideas from nowhere. They immerse themselves in activities that might generate ideas, such as reading science magazines, and they keep their antennae tuned for the faintest hint of an idea.

Mathematicians get ideas the same way. They read the math journals, they think about applications, and they keep their antennae tuned to "high."

Still, the very best seem to have other ways of thinking fresh thoughts. It's almost as if they lived on another planet. Srinivasa Ramanujan was a brilliant self-taught Indian mathematician whose life story is very romantic; it is well told in Robert Kanigel's *The Man Who Knew Infinity*. I prefer to think of Ramanujan as Formula Man. He learned most of his early mathematics from a single, rather curious textbook, George Carr's *A Synopsis of Elementary Results in Pure and Applied Mathematics*. It was a list of about five thousand mathematical formulas, starting with simple algebra and leading into complicated integrals in calculus and the summation of infinite series. The book must have appealed to Ramanujan's turn of mind, or he would never have worked his way through it; on the other hand, it led him to think (because he had no one to tell him otherwise) that the essence of mathematics is the derivation of formulas.

There is more to math than that: proof, for a start, and conceptual structure. But new formulas play a part, and Ramanujan was a wizard at them. He came to the attention of Western mathematicians in 1913 when he sent a list of some of his formulas to Hardy. Looking at this list, Hardy saw some formulas he could recognize as known results, but many others were so strange that he had no idea where they could have come from. The man was either a crackpot or a genius; Hardy and his colleague John Littlewood retired to a quiet room with the list, determined not to come out until they had decided which.

The verdict was "genius," and Ramanujan was eventually brought to Cambridge, where he collaborated with Hardy and Littlewood. He died young, of tuberculosis, and he left a series of notebooks that even today are a treasure trove of new formulas.

When asked where his formulas came from, Ramanujan replied that the Hindu goddess Namagiri came to him in dreams, and told them to him. He had grown up in the shadow of the Sarangapani temple, and Namagiri was his family deity. As I told you in an earlier letter, Hadamard and Poincaré emphasized the crucial role of the subconscious mind in the discovery of new mathematics. I think Ramanujan's dreams of Namagiri were surface traces of the hidden activity of his subconscious.

One can't aspire to be a Ramanujan. His kind of talent is uncanny; I suspect that the only way to understand it is to possess it, and even then it probably yields very little to introspection.

As a contrast, let me try to describe how I usually get new ideas, which is far more prosaic. I read a lot, often in fields unrelated to my own, and my best ideas often come when something I have read reminds me of something I already know about. That was how I came to work on animal locomotion.

The origin of this particular set of ideas goes back to 1983, when I spent a year in Houston working with Marty Golubitsky. We developed a general theory of space-time patterns in periodic dynamics. That is, we looked at systems whose behavior over time repeats the same sequence over and over again. The simplest example is a pendulum, which swings periodically from left to right and right to left. If you place a pendulum next to a mirror, the reflected version looks exactly the same as the original, but with one difference: when the reflection is at its extreme right position, the original is at its extreme left. These two states both occur in the original system, but there, they are separated by a time lag of exactly half the period. So the swinging pendulum has a kind of symmetry, in which a spatial change (reflect left–right) is equivalent to a temporal one (wait half a period). These space-time symmetries are fundamental to patterns in periodic systems.

We looked for applications of our ideas, and mostly found them in physics. For example, they organize and explain a host of patterns found in a fluid confined between two rotating cylinders. In 1985 we both went to a conference in Arcata, in northern California. After the conference was over, four of us—three mathematicians and a physicist—shared a rented car back to San Francisco. It was a very small car—calling it "subcompact" would be far too generous—and it had to hold all of our luggage as well as us. To make matters worse, we stopped off at a Napa Valley chateau so that Marty could pick up a crate of his favorite wine.

Anyway, on the journey we stopped every so often to admire the redwoods and giant sequoias, and in between Marty and I worked out how our theory applied to a system of oscillators joined together in a ring. ("Oscillator" is just a word for anything that undergoes periodic behavior.) We did the work entirely in our heads, not writing anything down because there wasn't room to move. This exercise was mathematically pleasing, but it seemed rather artificial. It never occurred to us to look at biology instead of physics, probably because we didn't know any biology.

At this point fate intervened. I was sent a book called *Natural Computation* to review for the magazine *New Scientist*. It was about engineers taking inspiration from nature, trying to develop computer vision by analogy with the eye, for instance. A couple of chapters were about legged locomotion: building robots with legs to move over rough terrain, that kind of thing. And in those chapters I came across a list of patterns in the movement of four-legged animals.

I recognized some of the patterns: they were space-time symmetries, and I knew that the natural place for them to occur was in a ring of four oscillators. Four legs . . . four oscillators . . . it definitely seemed promising. So I mentioned this curiosity in the review.

A few days after the book review appeared in print, my phone rang. It was Jim Collins, then a young research student visiting Oxford University, about fifty miles from where I lived. He knew a lot about animal movement, and was intrigued by the possible mathematical connection. He came to visit for a day, we put our heads together . . . to cut a long story short, we wrote a series of papers on space-time patterns in animal locomotion.

Many of the more radical changes of research direction in my life have come about in similar ways: spotting a possible connection between some math that I already knew and something I happened on by accident. Every link of this type is a potential research program, and the great beauty of it is, you have a pretty good idea how to get started. What features are crucial to the math? How might similar features appear in the real-world application? For example, in the locomotion story, the ring of mathematical oscillators relates to what neuroscientists call a "central pattern generator." This is a circuit made from nerve cells that spontaneously produces the natural "rhythms," the space-time patterns, of locomotion. So Jim and I quickly realized that we were trying to model a central pattern generator, and that a first stab was to treat it as a ring of nerve cells.

We no longer believe our original model is correct: it is too simple, it has a technical flaw, and something slightly more complicated is needed. We have a fair idea of what that replacement looks like. That's how research is: one good idea, and you're set for years.

Read widely, keep your mind active, keep your antennae out; when they report something interesting, pounce. As Louis Pasteur famously said, chance favors the prepared mind.

17

■ How to Teach Math

Dear Meg,

Excellent news! Congratulations on the postdoc position. I'm delighted, though not surprised: you deserve it. The research project on visual processing in fruit flies sounds very interesting, and it overlaps your interests pretty well, even if you haven't been exposed to the biological aspects before.

You should count it as a bonus that the position includes some teaching duties. You'll find that teaching math to others improves your own understanding. But it's only natural to be a little nervous, and I'm not surprised that you think you are "not at all prepared" for your teaching responsibilities. Many people in your position feel that way. But the nerves will vanish as soon as you get started. You've been in classrooms all your life, you've observed several dozen teachers at length, and you have strong opinions about what makes a course good or bad. All this is preparation. It's important that you not let a lack of confidence delude you into taking this part of your job too lightly.

A good teacher, like my Mr. Radford, is worth his or her weight in gold. Good teachers inspire their students, well, some of them. Correspondingly, bad teachers can put students off a subject for life. Unfortunately, it is much easier to be a bad teacher than a good one, and you don't have to be *really* bad to have the same negative effect as someone who genuinely is totally awful. It is far easier to destroy someone's confidence than to help them regain it.

Teaching matters. It's not just a boring necessity that pays for the excitement of research. It is your opportunity to pass on your understanding of math to the next generation. Many good mathematicians enjoy their teaching, and work just as hard at it as they do at their research projects. They feel a great pride of ownership in their courses.

It is not unusual for a research idea to occur to you while you are preparing a course, or teaching it, or setting a test on it. I think this happens because your mind moves out of its usual "research" grooves when you are teaching, and you start asking new questions.

Here I must add a confession. It is now several years since I taught any undergraduates, because my position was changed in 1997 to allow more time for "public understanding of science" activities. Instead of the usual fifty percent teaching, fifty percent research, it became fifty percent research and fifty percent public lectures, radio, TV, magazines, newspapers, and popular science books. This has had one beneficial effect: I've been able to employ teaching techniques you seldom encounter in an undergraduate lecture. The most memorable of these was the day I brought a live tiger into the lecture theatre.

I was delivering the 1997 Christmas Lectures, on BBC television: five hours of popular science, filmed "as live" before an audience of about five hundred mostly young people. This lecture series was initiated by Michael Faraday in 1826, and I was the second mathematician to take part.

One of the five lectures was on symmetry and pattern formation, and I wanted to start with William Blake's poem—a cliché, but a good one nonetheless—whose opening stanza begins "Tyger! Tyger! burning bright," and ends with "Dare frame thy fearful symmetry." So, television being what it is, we decided to bring on a real tiger. How we found one is a story in its own right, but suffice it to say, we did. Nikka was a six-month-old tigress, and she entered the lecture theatre on the end of a chain held by two burly young men.

I have never heard an audience go so quiet, so quickly.

Nikka's role went beyond poetic metaphor. The symmetry I had in mind was her stripes, especially the regular rings along her elegant tail. She was a true star: she behaved beautifully, and we got exactly the footage we wanted.

I've never really been able to match that as a way to start a lecture.

What I lost by way of contact with undergraduates, then, I gained in contact with wildlife. The Mathematics Department didn't suffer, because it was allocated a replacement teaching position, but I missed my regular interactions with undergraduate students. I gained in other ways, of course; mostly, I could allocate my time as I wished, which was wonderful. I still supervise PhD students, so some of my teaching has continued, but you should bear in mind that some of what I say may be outdated. In my defense, I did teach regular undergraduate courses for twenty-eight years before that, including two years in the United States, so I have some knowledge of the American system and how it differs from the British. The similarities are more important than the differences, but I'll try to translate my experience into your context.

To my mind, the most important feature of good teachers is that they put themselves in the student's position. It's not just a matter of giving clear and accurate lectures and grading tests; the main objective is to help the student understand the material. Whether you are delivering a lecture or talking with students during office hours, you have to remember that what seems perfectly obvious and transparent to you may be mysterious and opaque to someone who has not encountered the ideas before.

I always tried to remind myself of that. When grading tests, it is so easy to start thinking, "I've taught them this stuff for twenty years now, and they *still* don't understand it." But each year brings new students, who encounter much the same difficulties as their predecessors, make the same mistakes, misunderstand the same things. It's not *their* fault that you've seen it all before.

It's actually to your advantage, Meg, that you haven't seen it all before. Consider yourself fortunate, and exploit that. The students will feel comfortable with you because you're close to their age, you've just been through the same mill that is now grinding them, and you haven't grown bored teaching the same course several times. I can still remember my first few lecture courses vividly; teaching was easier for me then than it became ten years later. After a while, you know too much, and there's the danger that you try to pass *all* of that knowledge and insight on to the students. Big mistake. They don't have the same perspective that you do. So the KISS principle applies: Keep It Simple, Stupid. Stick to the main points, and try not to digress if doing so requires the students to understand new ideas that are not in the syllabus, however fascinating and illuminating they may seem to you.

The American system is more straightforward than the British one in this regard. Typically, there is a set text and an agreed syllabus—right down to page numbers and specific paragraphs to be included or not—so the content is established and everyone knows it, or should. But there's still room for input from the teacher, and there's a delicate balance between helping the students by putting your own stamp on the material, and confusing them by introducing too many extraneous ideas.

So before telling them something that may be outside the text, you need to ask yourself, if I were a student, who knew the textbook up to this particular page but nothing beyond, what would help me understand the material better? And the key step in coming up with a good answer to that question is to make sure you understand the material yourself.

Let me give you an example. The details are less important than the approach, which applies to many similar situations.

At some point early in your teaching career, one of your precalculus students is going to ask you why "minus times minus gives plus." For instance, $(-3) \times (-5) = +15$, not -15. And even though this topic should have been beaten to death in high school, you are going to have to justify the standard mathematical convention.

The first thing is to admit that it *is* a convention. It may be the only choice that makes sense, but mathematicians could, if they wished, have insisted that $(-3) \times (-5) = -15$. The concept of multiplication would then have been different, and the usual laws of algebra would have been torn to shreds and thrown out the window, but hey! Old words often take on new meanings in new contexts, and there is nothing sacred about the laws of algebra.

There are two reasons why the standard convention is a good one: an external reason, to do with how mathematics models reality, and an internal one, to do with elegance.

The external reason convinces a lot of students. Think of numbers as representing money in the bank, with positive numbers being money you possess and negative numbers being debts to the bank. Thus -5 is a debt of $5, so $3 \times (-5)$ is three debts of $5, which clearly amount to a total debt of $15. So $3 \times (-5) = -15$, and no one seems bothered much about that. But what of $(-3) \times (-5)$? This is what you get when the bank *forgives* 3 debts of $5. If it does that, you *gain* $15. So $(-3) \times (-5) = +15$.

The only other choice any students ever advocate is -15, but that would leave you in debt.

The "internal" explanation is to work out a sum like $(-3) \times (5 - 5)$. On the one hand, this is clearly zero. On the other, we can use the laws of algebra to expand it, getting $(-3) \times 5 + (-3) \times (-5)$. Since we've already agreed that minus times plus is minus, we deduce that consistency of the laws of algebra requires $-15 + (-3) \times (-5) = 0$, which implies that $(-3) \times (-5) = 15$ (add 15 to each side).

The most we can assert in the first case is that *if* we want the math to model bank accounts, then minus times minus has to work out as plus. The most we can assert in the second case is that *if* we want the usual laws of algebra to hold for negative numbers, then the same goes. There is nothing requiring either of these things to be true. But it will certainly be more convenient if they are, and that is why mathematicians chose that particular convention.

I'm sure you can think of other, similar arguments. The important thing is not to say to the student, "That's how it is. Don't question it, just learn it." But to my mind it would be even worse to leave them with the impression that there was never any choice to be made, that it is somehow *ordained* that minus times minus makes plus. All of those concepts—plus, minus, times— are human inventions.

At this point one of your more thoughtful students may remark that it *is* ordained, in the sense that there exists only one system of mathematics that proceeds from the counting numbers in a completely consistent way. You can respond that actually there are several extensions of the number concept—negative numbers, fractions, "real" numbers (infinite decimals), "complex" numbers in which -1 has a square root . . . even quaternions (in which -1 has many square roots but some laws of algebra fail). Each of these extensions is provably unique, subject to possessing certain features, but it is up to human beings to choose which features they consider significant. It would, for example, be possible to invent a new number system in which all negative numbers are equal, and it would be entirely consistent from a logical point of view. But it would not obey the usual rules of algebra.

You can concede, if pushed, that some extensions seem to be more natural than others.

Discussions of this kind do not always work; ingrained misconceptions can be hard to eradicate. Even if they do work, you need to show your students where their intuition is going wrong.

Typically, when a student gets stuck at some point in the syllabus, the real problem lies elsewhere, some pages or courses or years back. Perhaps they don't understand the relation between multiplication and repeated addition. Perhaps they understand it only too well, and can't see how you can add minus three lots of -5 together. It's amazing how often teaching reveals hidden assumptions or unquestioned features of your own mathematical background. Whenever an existing mathematical concept is extended into a new domain, you have to abandon some of its previous interpretations and accept new ones.

You can survive new material in mathematics for quite a while without taking it fully onboard. My colleague David Tall has a theory about this, which I rather like. His idea is that mathematics advances by (conceptually) turning processes into things. For instance, "number" starts out as the process of counting. The number 5 is where you get to when you count the fingers (by which we'll include the thumb) of one hand: "One, two, three, four, five." But in order to make progress, at some stage you have to stop going through the process of counting, and think of 5 as a *thing* in its own right. This is already useful when you start doing sums like 5 + 3. But students can cover up their inability to turn counting into a thing by lining up the fingers of one hand with three fingers of the other, and then counting the lot: "One, two, three, four, five, *six, seven, eight*."

This cover-up can go undetected for a long time, but it falls down in the face of sums like 2546 + 9773.

Multiplication affords another example. For a time, you can think of, say, 4 × 5 as 4 + 4 + 4 + 4 + 4 and fall back on your understanding of addition (even using counting). But when faced with 444 × 555, you need something more sophisticated.

The interesting thing is that the strategies that fail in the long term are precisely the ones that we use to teach these new concepts in the short term. We relate numbers to counting, often using real "counters." We relate multiplication to repeated addition. There's nothing wrong with this. Mathematics builds new ideas on old ones. It's hard to see how else you could teach it. But *eventually* the training wheels have to come off the bike: students have to internalize the new idea.

David calls these process-cum-concepts "procepts." A procept can sometimes be usefully viewed as a process, and other times as a concept, a thing. The art of mathematics involves switching effortlessly from one of those viewpoints to the other. When you're doing your research, you don't even notice the switch. But when you're teaching, you have to be aware of it. If one of your students is having trouble with a new procept, the cause may be a past failure to "proceptualize" one of the processes involved. So your job as a teacher is to backtrack through the series of ideas that leads up to the new one. You're not looking for the first place where your student fails to answer a question. You're looking for the first place where they can answer it only by using some simpler idea as a crutch.

In British elementary schools, the educational establishment has managed to get this spectacularly wrong. We now have a highly prescriptive "national curriculum," and teachers—quite literally—check hundreds of boxes to mark the student's progress. Can they count to five? Check. Can they add five to three? Check. The assumption is that what matters is their ability to get the answer. But what really matters is *how* they get the answer. I'm old-fashioned enough to believe that either way they have to get the *right* answer; no easy grades for "method" from me. But I am absolutely certain that checking a series of boxes is not the way to teach anyone mathematics.

18

■ The Mathematical Community

Dear Meg,

Now that you are on the verge of becoming a fully fledged member of the mathematical community, it's a good idea to understand what that entails. Not just the professional aspects, which we've already discussed, but the people you will be working alongside, and how you will fit in.

There's a saying in science fiction circles: "It is a proud and lonely thing to be a fan." The rest of the world cannot appreciate your enthusiasm for what seems to them a bizarre and pointless activity. The word "nerd" comes to mind. But we are all nerds about something, unless we are couch potatoes who have no interests except what's on TV. Mathematicians are passionate about their subject, and proud to belong to a mathematical community whose tentacles stretch far and wide. You will find that community to be a constant source of encouragement and support—not to mention criticism and advice. Yes, there will be disagreements too, but generally speaking, mathematicians are friendly and relaxed, provided you avoid pushing the wrong buttons.

Pride is one thing, loneliness another. My experience is that today's public is much more aware than it used to be that mathematicians do useful and interesting things. At parties, if you admit to being one, you are far more likely to be asked, "What do you think about chaos theory?" than be told, "I was never any good at math when I was at school." In *Jurassic Park*, Michael Crichton says that today's mathematicians no longer resemble accountants, and some are more like rock stars.

If so, this is very bad news for rock stars.

Even if people ask you about chaos theory at parties, it is still unwise to explain your latest theorem on semicontinuous pseudometrics on Kähler manifolds to the guy in a leather jacket. (Though nowadays he *might* turn out to be a mathematician. But don't count on it.) So, despite the public's newfound tolerance of math, there will be occasions when you want to be with people who understand where you're coming from. Such as just after you've finally proved the semicontinuous case of the Roddick–Federer conjecture on the irregularity of Kähler manifold pseudometrics in dimensions greater than 34.

Science fiction fans go to conventions ("cons," as they say) to talk to other science fiction fans. Whippet breeders go to whippet shows and compete with other people who breed whippets. Mathematicians go to conferences to hang out with other mathematicians. Or they give seminars, or colloquia, or just visit.

Our first vice chancellor, Jack Butterworth, once said that no university was worth anything unless a quarter of its faculty was in the air. He intended this literally: air travel, not intellectual high flight. The best way to advance the cause of mathematics is to meet other mathematicians.

If you are lucky, they will come to you. The University of Warwick, founded in the 1960s, became a world-class center for mathematics because from day one it held symposia, year-long special programs in some area of math. (I was once told that "symposium" means "drinking together," a theory that cannot be rejected out of hand.) But it's a good idea, and more fun, if you go to *them*. Mathematics, like all the sciences, has always been international. Isaac Newton used to write to his counterparts in France and Germany, but today he could hop on a budget flight and meet them.

Mathematicians get together, usually over coffee; Erdős said that a mathematician is a machine for turning coffee into theorems. They share jokes, gossip, theorems, and news.

The jokes are mathematical, of course. There is a lengthy compendium of classic mathematical jokes in the January 2005 issue of the *Notices of the American Mathematical Society*, and its contents are a vital part of your mathematical culture, Meg. There is, for instance, a Noah's ark joke. (Actually, my favorite Noah's ark joke is biological: a cartoon. The rain is coming down in sheets, the ark is loaded with two of every kind of animal, and Noah is on hands and knees grubbing around in the mud. Mrs. Noah is shouting from the ark, "Noah! Forget the other amoeba!") Anyway, the mathematical Noah's ark joke goes like this:

The Flood has receded and the ark is safely aground atop Mount Ararat; Noah tells all the animals to go forth and multiply. Soon the land is teeming with every kind of living creature in abundance, except for snakes. Noah wonders why. One morning two miserable snakes knock on the door of the ark with a complaint. "You haven't cut down any trees." Noah is puzzled, but does as they wish. Within a month, you can't walk a step without treading on baby snakes. With difficulty, he tracks down the two parents. "What was all that with the trees?" "Ah," says one of the snakes, "you didn't notice which species we are." Noah still looks blank. "We're adders, and we can only multiply using logs."

This joke is a multiple pun: you can multiply numbers by adding their logarithms. Other jokes parody the logic of proofs: "*Theorem*: A cat has nine tails. *Proof*: No cat has eight tails. A cat has one more tail than no cat. QED."

Mathematicians tell each other theorems. Quirky ones, like the "ham sandwich theorem": if you have a slice of ham and two slices of bread, arranged in space in any relative positions whatsoever, then there exists a plane dividing each of the three pieces exactly in half. Or the recently proved "bellows conjecture," which says that if a polyhedron flexes (as, remarkably, some can), then its volume doesn't change. But there is often a sting in the tail: "Proved that? OK, now do it with n objects in n dimensions." Sometimes they tell each other conjectures, theorems not yet proved and that for all they know might be false. My favorite is the "sausage conjecture." For starters, suppose you want to wrap a number of tennis balls in plastic film. What arrangement has the least surface area? (Assume that the film forms a convex surface: no dents.) The answer is that if you have fifty-six balls or fewer, they should be placed in a line to make a "sausage." If you have fifty-seven or more, then they should be clumped together more like potatoes in a string bag.

In a four-dimensional analogue, the breakpoint is somewhere between fifty thousand and one hundred thousand. With fifty thousand balls they form a sausage. With one hundred thousand they clump. The exact breakpoint here is not known.

Here is the full conjecture: Three and four dimensions are misleading. Prove that in five or more dimensions, sausages are *always* the answer, no matter how large the number of balls may be.

The sausage conjecture has been proved in forty-two dimensions or more.

This is bizarre. I love it.

There will be gossip. Nowadays it may be about the topologist who ran off with her secretary, or the messy divorce of two well-known group theorists, but that's a recent development that I trace to the bad influence of television. Traditionally, gossip is about who is in line for the Chair of Abstract Nonsense at Boondoggle University, or do you know anyone who has a postdoc position going for a young functional analyst like my student Kylie, or do you think Winkle and Whelk's purported proof of the mass gap hypothesis has any chance of being right?

There will be serious news. As I write, a major topic of conversation is the latest information on Grisha Perelman's alleged proof of the Poincaré conjecture. Has anyone found a hole in it yet? What do the experts think today? This is really exciting because the Poincaré conjecture is one of the great open questions in mathematics, second only to the Riemann hypothesis. It all went back to a mistake that Henri Poincaré made in 1900. He assumed without proof that any three-dimensional topological space (with some technical conditions) in which every loop can be continuously shrunk to a single point must be equivalent to a three-sphere, the three-dimensional analogue of the two-dimensional surface of an ordinary sphere. Then he noticed the absence of any proof, tried to find one, and failed. He turned the failure into a question: is every such space a three-sphere? But everyone was so sure that the answer had to be yes that his question quietly turned into a conjecture. Its generalization to higher dimensions was then proved, for every dimension *except* 3, which was disappointing, to say the least. The Poincaré conjecture became so notorious that it is now one of the seven millennium problems selected by the Clay Institute as the most important open questions in mathematics. Each problem carries a million-dollar reward for its solution.

In 2002 and 2003 Perelman, a rather diffident young Russian with a physics background, published two papers on the arXiv ("archive"), a website for mathematical preprints, with the offhand remark that they not only proved the Poincaré conjecture, they also proved the even more powerful Thurston geometrization conjecture, which holds the key to *all* three-dimensional topological spaces!

Usually this kind of claim turns out to be nonsense, but Perelman's idea is clever and comes with a good pedigree. His trick is to use the so-called Ricci flow to deform the candidate space in a manner closely analogous to how space-time deforms under Einstein's equations of general relativity. And that's the snag. To understand the proof properly, you need to know three-dimensional topology, relativity, cosmology, and a dozen other hitherto unconnected areas of pure math and mathematical physics. And it's a long and difficult proof, with plenty of traps for the unwary. Moreover, Perelman followed the time-honored Russian tradition of not giving all the details. So the experts, who have been working through his ideas in seminars all over the world, are understandably wary of declaring the proof correct. But every time someone finds what might be a gap or a mistake, Perelman quietly explains that he's already thought of that and why it isn't a problem. And he's right.

It's gotten to the point where, even if the proof turns out to be wrong, the correct things achieved along the way are of major significance to mathematics. And as I write, the experts seem to be nudging ever closer to the view that the proof really does work. Keep your ears open over the coffee, Meg.

As your career develops, the worldwide mathematical community will be increasingly important to you. You will become part of it, and then you will have a home in every city on Earth.

Just arrived in Tokyo? Drop by the nearest university, find the math department, walk in. There will be at least one person you know, or who knows you by your work even if you've not met before. They will drop everything, call their baby-sitter, and take you out on the town for the evening. They may have a spare room, if you forgot to book a hotel. They will set up a seminar so that you can present your latest ideas to a sympathetic audience.

They may even be able to drum up a small financial contribution to your airfare.

You don't get to fly business class, though. Or to sleep in a hotel suite. (Not *yours*, at least.) Math operates on the cheap and cheerful principle. I sometimes wish we didn't undervalue ourselves in this manner, but it's ingrained habit and it is far too late to change it.

It is of course more civilized and more organized to e-mail the University of Tokyo math department in advance. The result will be similar.

If you get on well with your host, they will invite you back. As you and they climb the career ladder, both of you will start being invited to conferences. Then you will find yourself organizing conferences, which means that you can invite everyone you want to talk to. There is some kind of "phase transition," so that over a period of about a year, you will go from being invited to no conferences to being invited to far too many. Be selective; learn to say no. Learn sometimes to say yes.

There are big conferences and medium-sized conferences and small conferences. There are special conferences and general ones. The big, general ones are great for meeting people and trawling for jobs. Every four years, the International Congress of Mathematicians is held somewhere in the world. I last went when it was in Kyoto, and there were four thousand participants. I saw a lot of Kyoto, met lots of old friends and made some new ones, and learned a little bit about what people outside my area were doing. The family came too, and they had a whale of a time exploring the city and its surroundings.

I much prefer smallish, specialized meetings with a specific research theme. You can learn a lot from those, because almost every talk is on something that interests you and is related to what you are working on at the moment. And once you've been in the business for a few years, you will know almost everyone else who is attending. Except for the youngest participants, who have only just joined the community.

Welcome, Meg.

19

■ Pigs and Pickup Trucks

Dear Meg,

Assistant Professor, indeed. I'm proud of you; we all are. At an excellent institution, too. You're a professional mathematician now, with professional obligations. And it occurs to me that I've been so busy offering advice about what to do in various circumstances that I've left out the other side of the equation: what not to do. Now that you have a tenure-track position, you will be taking on more responsibility, so you will have more to lose if you foul up. There are plenty of ways for mathematicians to make complete idiots of themselves in public, and nearly all of us have managed it at some stage in our careers. People make mistakes; wise people learn from them. And the least painful way is to learn from mistakes made by others.

The longer you stay in the business of mathematics, the more blunders you will inevitably make; this is how experienced people gain their experience. I have witnessed, and committed, plenty of mistakes myself. They can range from writing the wrong equation on the board to mortally insulting the president of your university at some significant public event. Be warned. You will no doubt invent some new mistakes of your own; occasional embarrassment is the natural human condition.

Most of my advice will be obvious. An assistant professor who wishes to be tenured at her university must find out what the requirements and expectations are, and then meet them. If you are expected to have published two papers beyond the subject of your dissertation and instead publish one paper, coach the math club, direct the study-abroad program in Budapest, obtain a major research grant, and win the teacher-of-the-decade prize, you may be denied tenure. Take care to be polite to your superiors, unless you have excellent reasons not to and want to change jobs. Be polite to everyone else, when they deserve it and even sometimes when they don't. If you disagree with some decision or argument, make your point concisely, clearly, and without implying that the opposing view is insane, even when it is. Honor your commitments, whether they are tutorial sessions, office hours, examination grading, or plenary lectures at the International Congress of Mathematicians. If you agree to sit on a committee, turn up for its meetings. Listen to the discussion. Contribute, though not at length. Generally, remember that you are a professional, and behave like one.

Some mistakes, on the other hand, are obvious only after you've made them. There is a persistent story at Warwick University that I ended my first ever undergraduate lecture by walking into the broom cupboard. It is time to set the record straight. Yes, I admit that it *was* a broom cupboard, but it was also the emergency exit from the lecture hall. I had assumed, without finding out ahead of time, that when the students left the hall by the main doors, I would be able to leave by what looked like a side door. But when I tried it, I found myself surrounded by buckets and mops. Worse, I discovered that the only way to leave the building by that route was to push open an emergency exit, which would set off an alarm. I had noticed the EXIT sign over the door but had failed to spot the word "emergency" above it. So I was forced, rather sheepishly, to emerge from the so-called broom cupboard and join the students as they walked up the stairs to the back of the hall and out the main doors.

The message here is, don't assume things. Check them beforehand. Not just the layout of the lecture hall or the location of the building where you are supposed to be giving a talk, or the city in which your meeting is due to take place, or the date of that meeting . . . Recall Murphy's law: "Anything that can go wrong, will." Above all, remember the mathematician's corollary to Murphy's law: "Anything that *can't* go wrong will go wrong too."

This was brought home very clearly to a good friend of mine, also a mathematician, on a trip to a country that it would be best not to name. He was attending a conference and he was making a flight from his first port of entry to another, fairly distant, city. As he sat in the airplane scribbling some calculations, he noticed the pilot emerge from the flight cabin, shutting the connecting door behind him. A few minutes later the copilot did the same thing. Soon after that, the pilot returned, and tried to open the door to return to the cockpit. He seemed to be encountering some difficulty. The copilot tried to help, but neither of them could open the door. At this point my friend realized that the plane was flying on autopilot and no one could get to the controls. A female flight attendant joined the pilot and copilot, disappeared, and reappeared carrying a small hand ax. The pilot then attacked the door with the ax, made a hole in it, put his hand through, and opened the door. The flight crew then entered the cabin and closed the door behind them.

No announcement was made to explain these events to the bemused and distinctly frightened passengers.

My advice here really applies to the pilot and the copilot, not the passengers. If you go to conferences, you will sometimes have to fly on airlines that have been booked by the conference organizers. You can choose not to go, if you wish, but you can't always choose not to travel on an airline with a dubious safety record or an aging fleet.

There really is no way for a passenger to anticipate such a problem, or to help solve it or avoid it.

Let me revert to the topic of lecturing. Another useful piece of advice is to make sure that you have plenty of time to get to the lecture room. Avoid taking on unpredictable commitments immediately beforehand. I still have vivid recollections of arriving late to give a lecture on algebra. I lived in a village at that time, and another member of the math department owned a small farm in the same village, so we carpooled. One day when it was his turn to drive, he decided to drop a pig off at the local abattoir on the way to work. The pig, perhaps sensing that the trip would not be to its advantage, had other ideas. It refused to climb the plank into the back of the truck. It is difficult to remain professorial when explaining that you were late because you could not get a pig into a pickup truck.

One of the main ways you will interact with other mathematicians is by attending, and giving, talks. These might be seminars, specialist talks for experts in your research area; they might be colloquia, more general talks for professional mathematicians but not specialists in the area concerned; or they might be public lectures open to anyone who wants to turn up. All lectures are fraught with potential disaster.

There was, for instance, a prominent professor of number theory who had the habit of turning up at the start of a visiting speaker's seminar, falling asleep within minutes of the talk starting, snoring loudly the whole way through, and then asking penetrating questions at the end when the audience's applause woke him up. By all means, emulate the penetrating questions, but try to avoid the snoring if you possibly can; otherwise you will get a reputation for eccentricity.

If you are the person giving the talk, the opportunities for Murphy to strike are far more numerous, especially if you are using equipment. When I started lecturing, the only equipment we ever had was blackboard and chalk, and I confess to a continuing bias in favor of lo-tech visual aids, though I am capable of preparing the all-singing, all-dancing PowerPoint presentation with a video projector and moving graphics downloaded from the Internet if that is what is called for. I have used overhead projectors, whiteboards with those horrible pens that smell of solvent, even the businessperson's ubiquitous flipchart.

Even chalk can go wrong. First, it may not be there. I developed a habit of taking my own box of chalk to lectures in case the previous lecturer had used it all, or the students had hidden it as a joke. Some chalk is very dusty and gets all over your clothes: I made sure I had the "antidust" kind with me. It still got on my clothes, but the cleaning bills were smaller. Many types of chalk can make horrible screeching sounds when you write, putting everyone's teeth on edge. It takes practice to prevent that. And normal-sized chalk will not do if you are lecturing to five hundred freshman calculus students in a cavernous lecture hall. You need supersize.

Other types of equipment can go wrong more spectacularly. A good friend of mine was delivering a short talk at the British Mathematical Colloquium—the UK's main mathematics conference, held annually—and he was intending to use an overhead projector to show lots of pictures on a projection screen. Unfortunately, when he tried to pull the screen down from the ceiling—it was on a roller and there was a string attached—it fell down on his head. He ended up projecting the pictures onto the wall.

Never believe your hosts when they tell you that all the equipment will work perfectly. Always try it for yourself before the lecture. I was giving a public lecture in Warsaw using a cassette of about eighty 35-mm slides. I was persuaded to hand the slides to the projectionist, who would set everything up for me, while my hosts took me for a quick coffee. As I entered the room to deliver the lecture, the projectionist put the slides into the projector, which was tilted at an alarming angle because of the high position of the screen. The cassette slid right through the projector and fell out the back onto the floor, where its contents were scattered far and wide. Many of the slides were sandwiched between thin sheets of glass, which broke. It took ten minutes to get everything back into some kind of order, in front of five hundred patient people.

Do not confuse the projection screen with a white-board and write on it in permanent ink. Many people do this, and their lecture is preserved for eternity, along with their mistake.

The famous physicist Richard Feynman once learned Spanish because he was going to give a lecture in Brazil. Check the local language.

If you are hosting a visitor, and they are giving a talk, make sure that you have the key to the projection room. On one occasion I had to improvise my lecture because the slide projector, though visible to us all inside a wonderfully equipped room, might as well have been located on the moon because the room was locked and no one knew how to get the key.

Do not forget that your visitor may not know the local geography. I was once abandoned inside a Dutch mathematics department building when my hosts went off to the parking garage to go to a restaurant. I had to make my escape through a window, setting off a burglar alarm. I did manage to catch up with them in the garage, which was a good thing because I had no idea where the restaurant was, or even what its name was.

Avoid wandering around strange buildings, especially in the dark. A biologist friend was visiting an institution with a strong marine biology department, and it had an aquarium. The entrance was down a short flight of stairs. Alone in the building, late at night, he tried to enter the

pitch-black room, felt for the light switch, and acciden-
tally hit the fire alarm instead.

Six fire engines turned up, lights blazing, sirens
blaring.

He had called the fire station to explain his mistake,
but according to the rules, they could not return to base
without checking for a fire.

Committees are another place where you can easily
make dreadful mistakes. Universities tend to run as a
network of interlocking committees and subcommittees,
some with real teeth, some window dressing. Most exist
for a reason, and many tackle low-level but essential activities such as grading tests or sorting out course regula-
tions. You will undoubtedly be involved in committee
work, and you should be. A university is a complicated
place, and it will not function well if everyone is left to
make things up as they go along. Every academic has to
be something of an administrator, and to some extent
the converse holds, especially at senior levels.

Not being a committee animal myself, I can't offer
much useful advice on how to "work" a committee so
that it reaches the decision you favor. But I do know how
not to. The following story is typical. An important
committee was debating a major decision about some
particular action. The mathematician on the committee
saw immediately that taking the action concerned would
lead to disaster, and spent five minutes explaining the
logic behind this view, which was unassailable. His

analysis was clear and concise, and left no real room to doubt his conclusions; no one contradicted him. The debate continued, however, because the other members of the committee had not yet had their say. After an hour or more of further discussion—to which the mathematician did not contribute, having already (he assumed) made his point—the committee voted. It decided to take precisely the action that the mathematician had warned it not to.

What was his mistake? It was not in the analysis, nor in its presentation, but in its timing. In any committee discussion, there comes a pivotal moment when the decision can be swayed either way. That is the time to strike. If you make your point too soon, everyone else will have forgotten it; if you are lucky, you may be able to remedy this with a timely reminder. But if you make your point too late, there is no way it can have any effect.

The other thing not to do in committees is to keep making your point when you have already won it. You may lose support merely because you keep banging on about what has by now become obvious to everyone. If you have further ammunition, save it in case it is needed later.

Which advice I will now take myself.

20

■ Pleasures and Perils of Collaboration

Dear Meg,

Yes, it is a bit of a dilemma. Tenure and promotion depend on your own personal record of teaching and research, but there are attractions in working with others as part of a team. Fortunately, the advantages of collaborative research are becoming widely recognized, and any contribution that you make to a team effort will be recognized too. So I think you should focus on doing the best research you can, and if that leads you to join a team, so be it. If the research is good, and your teaching record is up to scratch, then promotion will follow, and it won't matter whether you did the work on your own or as part of a joint effort. In fact, collaboration has definite advantages; for instance, it's a very effective way to get experience in writing grant proposals and managing grants. You can start out as a junior member of someone else's team, and before long you will be a principal investigator in your own right.

Attitudes are changing, fast. In the past, mathematics was mostly a solo activity. The great theorems were discovered and proved by one person, working alone. To be sure, their work was carried out alongside that of other (equally solitary) mathematicians, but collaborations were rare and papers by three or more people were virtually nonexistent. Today, it is entirely normal to find papers written by three or four mathematicians. Something close to ninety-eight percent of my research in the last twenty years has been collaborative; my record is a paper with nine authors.

This is small potatoes compared to other branches of science. Some physics papers have well over a hundred authors, as do some papers in biology. The growth of collaboration has sometimes been derided as an adaptation to the "publish or perish" mentality, whereby tenure and promotion are determined by how many papers someone has published in a given period of time. An easy way to increase your list of publications is to get yourself added to somebody else's paper, and the payback is simple: you add them to yours.

But I really don't think this tit-for-tat behavior is responsible, to any significant degree, for the growth of coauthorship.

The reason for the huge cast of authors on some physics papers is straightforward. In fundamental particle physics, the joint work of a huge team, carried out over several years, is typically condensed into a paper of perhaps four journal pages. The team will include theorists, programmers, experts in the construction of particle detectors, experts in pattern recognition algorithms that interpret the extremely complex data obtained by the detectors, engineers who know how to construct low-temperature electromagnets, and many others. All of them are vital to the enterprise; all fully deserve recognition as authors of the resulting report. But the report is usually short and to the point. "We have detected the omega-minus particle predicted by theory: here is the evidence." Someone may get a Nobel Prize for those four pages. Probably whoever is listed first.

Big science involves big numbers of people. The same goes for gigantic biology projects such as genome sequencing.

Something similar has been going on throughout science. The root cause is that science and mathematics have become increasingly interdisciplinary. For example, you'll recall that one area of interest to me is the application of dynamics to animal movement, and my early papers were all written in collaboration with Jim Collins, an expert in biomechanics. They had to be: I didn't know enough about animal locomotion and Jim wasn't familiar with the relevant mathematics.

My nine-author paper summed up two three-year projects to apply new methods of data analysis to the spring and wire industries. More than thirty people were involved in this work; for publication we reduced the list of authors to those who had been directly responsible for a significant aspect of the results. Some of us worked on theoretical aspects of the mathematics; some worked out new ways to extract what we needed from the data we could record; others carried out the analysis of those data. Our engineers designed and built test equipment; our programmers wrote code so that a computer could perform the necessary analysis in real time. Interdisciplinary projects are like that.

For the last twenty years, funding bodies worldwide have advocated the growth of interdisciplinary research, and rightly so, because this is where many of the big advances are being made, and will continue to be made. At first they didn't get it quite right. The idea of interdisciplinary research was praised, but whenever anyone put in a proposal for such work, it went to the existing single-discipline committees, who of course did not understand substantial parts of the grant application. For example, a proposal to apply nonlinear dynamics to evolutionary biology would be turned down by the mathematics committee because they and their referees had no expertise in evolution, and then it would be rejected by the biology committee because they didn't understand the math. The result was that the funding agencies encouraged interdisciplinary research in every way except kicking in funds.

No one, mind you, was doing anything *wrong*. It was virtually impossible to justify a decision to spend money on the dynamics of evolution when that took money away from top-class projects about algebraic topology or protein folding, say. To the credit of those involved, the system has changed, for the better, and one can now obtain funding for science and math that straddles several disciplines. An important consequence is the creation of entirely new disciplines, such as biomathematics and computational cosmology. Another is the blurring of traditional subject boundaries.

Setting these political factors aside, there is another reason why collaborative publication has increased dramatically. A social reason. Working in groups, with colleagues, is a lot more fun than sitting in your office with your computer. To be sure, sometimes you need solitude, to sort out a conceptual problem, formulate a definition, carry out a calculation. But you also need the stimulus of discussions with others in your own area, or in areas where you wish to apply your ideas. Other people know things that you don't. More interestingly, when two people put their heads together, they sometimes come up with ideas that neither of them could have had on their own. There is a synergy, a new synthesis, something that Jack Cohen and I like to call *complicity*. When two points of view complement each other, they don't just fit together like lock and key or strawberries and cream; they spawn completely new ideas. As your career progresses, you may well come to appreciate the delights of collaboration, and value the help, interest, and support of colleagues whose minds complement your own.

Unfortunately, there can sometimes be a downside to collaboration. Choosing the wrong collaborator is a surefire recipe for disaster. Unfortunately, this can happen with entirely reasonable and competent people; it's a matter of personal "chemistry," and not always easy to predict. The main thing is to be aware of the possibility, and to leave yourself an exit strategy.

Some years ago two mathematicians that I know were coauthoring a book. They had no trouble agreeing on the mathematical content, or the order in which the material was presented. They just couldn't agree on the punctuation. It got to a stage where one of them would go through the entire manuscript putting in commas, and then the other would take them all out again. And round and round it went. The book did get written, but they have never collaborated on another one. They remain the best of friends, though.

Everyone involved in a collaboration must bring something useful to it. They don't have to do the same amount of work; one person may be the only one who knows how to do a big calculation, or write a complicated program for the computer, while another may contribute a crucial idea that ends up as two lines in the middle of a proof. As long as everyone contributes something essential, that's fair. No one objects to all coauthors getting some of the credit for the final result.

But if one of the participants is just along for the ride, which occasionally happens, then it makes everyone else happier if their name does not appear on the final paper, book, or report. And it often makes the person concerned happier, too. This need not be a sign of laziness; sometimes the project changes direction in a way that could not have been anticipated, and a contribution that originally looked essential may turn out not to be needed.

In big science, where huge teams are involved, the project plan usually stays fairly rigid and people drop out only if they leave and are replaced. But mathematical collaborations tend to be loose and spontaneous, and if there is a project plan, the first item on the list is to be ready to change the plan.

It helps a lot to be relaxed and tolerant. That doesn't prevent arguments; quite the contrary. The best of friends can have long, loud, and emotionally heated disputes during a research project. Psychologists now think that the rational part of our brain rests on the emotional part: you have to be emotionally committed to rational thinking before you can think rationally. With some of my collaborators, neither of us feels the project is getting anywhere unless we have a shouting match every so often. But the shouting stops as soon as we both sort out who is right, and there is no residual resentment. We are relaxed about having an argument; we are not so relaxed that arguments do not happen.

Never enter into a collaboration merely because you have been convinced that you should. Unless you are genuinely interested in working with someone, don't. It doesn't matter how big an expert they are, or how much grant money the project would bring in. Stay away from things that do not interest you.

On the other hand, I do find that it pays to have broad interests. That way, the list of things you should stay away from is much smaller. I once had a fascinating lunch with a medievalist who was an expert on the use of commas in the Middle Ages. Nothing came of that interaction, but it does occur to me that my two friends might have benefited from his presence on their book-writing team.

21

■ Is God a Mathematician?

Dear Meg,

It was very good to see you in San Diego last month. I'm ashamed to say I'd rather lost touch with your parents since they moved to the country. I wrote to them and was glad to hear your dad is on the mend.

People react to getting tenure in interesting ways. Most continue their teaching and research exactly as before, but with reduced stress. (This does not apply in the UK, by the way, because tenure was abolished there twenty years ago.) But I do remember one colleague who earnestly declared his intention to publish no more than one paper every five years for the rest of his career. That, he said, was the frequency with which good ideas came to him. It was an honest attitude, but possibly not a wise one. Another devoted himself almost exclusively to consulting work; within two years he'd left the university to start his own company. He now has a vacation home on one of the Caribbean islands. Apparently he got tired of "cheap and cheerful."

You, I see, have reacted by growing philosophical.

The physicist Ernest Rutherford used to say that when a young researcher in his lab started talking about "the universe," he put a stop to it immediately. I'm more relaxed about such talk than Rutherford was. My main reservation is that the territory should not be reserved solely for philosophers.

Two and a half thousand years ago, Plato declared that God is a geometer. In 1939 Paul Dirac echoed this, saying, "God is a mathematician." Arthur Eddington went a step further and declared God to be a *pure* mathematician. It is certainly curious that so many philosophers and scientists have been convinced of a fundamental link between God and mathematics. (Erdős, who thought God had other fish to fry, still believed He kept a Book of Proofs close at hand.)

God and mathematics both strike terror into the heart of common humanity, but the connection must surely run deeper. This is not a question of religion. You needn't subscribe to a personal deity to be awestruck by the astonishing patterns in the universe or to observe that they seem to be mathematical. Every spiral snail shell or circular ripple on a pond shouts that message at us.

From here it's a short step to seeing mathematics as the fabric of natural law and dramatizing that view by attributing mathematical abilities to a metaphorical or actual deity. But what *are* laws of nature? Are they deep truths about the world, or simplifications imposed on nature's unutterable complexity by humanity's limited brainpower? Is God really a geometer? Are mathematical patterns really present in nature, or do we invent them? Or, if real, are they merely a superficial aspect of nature that we fixate on because it's what we can comprehend?

The reason we cannot answer these questions definitively is that we human beings cannot step outside ourselves to obtain an objective view of the universe. Everything we experience is mediated by our brains. Even our vivid impression that the world is "out there" is a wonderful trick. The nerve cells in our brains create a simplified copy of reality inside our heads and then persuade us that we live inside it, rather than the other way around. After hundreds of millions of years of evolution, the human brain's abilities have been selected not for "objectivity" but to improve its owner's chances of survival in a complex environment. As a result, the brain is not at all a passive observer of nature. Our visual system, for example, creates the illusion of a seamless world that envelops us completely, yet at any instant our brains are detecting only a tiny part of the visual field.

Because we cannot experience the universe objectively, we sometimes see patterns that do not exist. About two thousand years ago, one of the strongest pieces of evidence for the existence of a geometer God was the Ptolemaic theory of epicycles. The motion of every planet in the solar system was held to be built up from an intricate system of revolving spheres. How much more mathematical can you get? But appearances are deceptive, and today this system strikes us as nonsensical and overly complex. It can be adjusted to model any kind of orbit, even a square one. Ultimately it fails, because it cannot lead us to an explanation of *why* the world should be this way.

Compare Ptolemy's wheels within wheels to Isaac Newton's clockwork universe, set in motion at the moment of creation and thereafter obeying fixed and immutable mathematical rules. For example, the acceleration of a body is the force acting on it divided by its mass. This one law explains all kinds of motion, from cannonballs to the cosmos. It has been refined to take relativistic and quantum effects into account in the realms of the very small or the enormously fast, but it unifies an enormous body of observational evidence. The tiny ripples discovered recently in the cosmic microwave background show that when the big bang went off, the universe did not explode equally in all directions. This asymmetry is responsible for the clumping of matter without which you and I wouldn't have a leg—or a planet—to stand on. It's an impressive verification of the modern extensions of Newton's laws, and it shows that patterns need not be perfect to be important.

It is no coincidence that Newton's laws deal with forms of matter and energy that are accessible to our senses, such as force. If we ride on a fairground roller-coaster, we feel ourselves pulled off our seats as the vehicle careers over a bump. But again our brains are playing tricks. Our senses do not react directly to forces. In our ears are devices, the semicircular canals, that detect not force but acceleration. Our brains then run Newton's law in reverse to provide a sensation of force. Newton was "deconstructing" his sensory apparatus back into the laws that made it work to begin with. If Newton's laws hadn't worked, then his ears wouldn't have worked either.

We have gotten much better at spotting the artificiality of putative patterns like Ptolemy's, systematic delusions created by a mathematics that is so adaptable that it can explain anything. One way to eliminate these delusions is to favor simplicity and elegance: Dirac's provocative point, and the true message of Occam's razor.

One of the simplest and most elegant sources of mathematical pattern in nature is symmetry.

Symmetry is all around us. We ourselves are bilaterally symmetric: we still look like people when viewed in a mirror. The symmetry is not perfect—normally hearts are on the left—but an almost-symmetry is just as striking as an exact one, and equally in need of explanation. There are precisely 230 symmetry types of crystals. Snowflakes are hexagonally symmetric. Many viruses have the symmetry of a dodecahedron, a regular solid made from twelve pentagons. A frog begins life as a spherically symmetric egg and ends it as a bilaterally symmetric adult. There are symmetries in the structure of the atom and the swirl of galaxies.

Where do nature's symmetric patterns come from? Symmetry is the repetition of identical units. The main source of identical units is matter. Matter is composed of tiny subatomic particles, and all particles of a given type are identical. All electrons are exactly the same. The famous physicist Richard Feynman once suggested that perhaps there is only one electron, batting backward and forward in time, and we observe it multiple times. Be that as it may, the interchangeability of electrons implies that potentially the universe has an enormous amount of symmetry. There are many ways to move the universe and leave it looking the same. The symmetries of a spiral snail shell or the drops of dew spaced along a spider's web at dawn can be traced back to this pattern-forming potential of fundamental particles. The patterns that we experience on a human scale are traces of deeper patterns in the structure of space-time.

Unless, of course, those deeper symmetries are only imaginary, the modern version of epicycles.

That the universe we experience is a contrivance of our imaginations, however, does not imply that the universe itself has no independent existence. Imagination is an activity of brains, which are made from the same kind of materials as the rest of the cosmos. Philosophers may debate whether the pattern that we detect in a tiger's stripes is really present in an actual tiger; but the pattern of neural activity evoked in our brains by the tiger's stripes is definitely present in an actual brain. Mathematics is an activity of brains, so they at least can on occasion function according to mathematical laws. And if brains really can do that, why not tigers too?

Our minds may indeed be just swirls of electrons in nerve cells; but those cells are part of the universe, they evolved within it, and they have been molded by Nature's deep love affair with symmetry. The swirls of electrons in our heads are not random, not arbitrary, and not—even in a godless universe, if that is what it is—an accident. They are patterns that have survived millions of years of Darwinian selection for congruence with reality. What better way to build simplified models of the world than to exploit simplicities that are actually there? Imaginary systems that get too far removed from reality are not useful for survival.

Intellectual constructs like epicycles or laws of motion may be either deep truths or clever delusions. The task of science is to provide a selection process for ideas that is just as stringent as that employed by evolution to weed out the unfit. Mathematics is one of its chief tools, because mathematics mimics the pictures in our heads that let us simplify the universe. But unlike those pictures, mathematical models can be transferred from one brain to another. Mathematics has thus become a crucial point of contact between different human minds; and with its aid, science has come down in favor of Newton and against Ptolemy. Even though Newton's laws—or, more to the point, their modern successors, relativity and quantum theory—may eventually turn out to be delusions, they are much more productive delusions than Ptolemy's.

Symmetry is a better delusion still. It is deep, elegant, and general. It is also a geometric concept. So the geometer God is really a God of symmetry.

Perhaps we have created a geometer God in our own image, but we have done it by exploiting the basic simplicities that nature supplied when our brains were evolving. Only a mathematical universe can develop brains that do mathematics. Only a geometer God can create a mind that has the capacity to delude itself that a geometer God exists.

In that sense, God *is* a mathematician; and She's a lot better at it than we are. Every so often, She lets us peek over her shoulder.

Notes and References

Page 21   and baffled by open questions like the Riemann hypothesis.   The Riemann hypothesis concerns Riemann's zeta function $\zeta(z)$, which makes it possible to convert questions about prime numbers into questions in complex analysis. It states that if $\zeta(z)$ is zero then either z is twice a negative integer or the real part of z is ½. See Karl Sabbagh, *Dr. Riemann's Zeros*, Atlantic Books, London 2002.

Page 23   the prevalent belief that the human sperm count is falling.   P. Bromwich, J. Cohen, I. Stewart, and A. Walker, Decline in sperm counts: An artefact of changed reference range of "normal"? *British Medical Journal* 309 (2 July 1994) 19–22.

Page 50   Leonardo of Pisa, also known as Fibonacci.   "Fibonacci" means "son of Bonaccio." This nickname was probably invented by Guillaume Libri in the nineteenth century, and certainly does not go back much earlier.

Page 57   the only plausible symmetric network that could explain all of the standard gaits of four-legged animals.   M. Golubitsky, I. Stewart, J. J. Collins, and P.-L.

Buono, Symmetry in locomotor central pattern generators and animal gaits, *Nature* 401 (1999) 693–695.

Page 84   *as some say the Bible does, but only if you take an obscure passage extremely literally.*   1 Kings 7:23 reads, "And he [Hiram on behalf of King Solomon] made a molten sea, ten cubits from the one brim to the other: it was round all about, and his height was five cubits: and a line of thirty cubits did compass it round about." If we assume the geometry is a circle, and if we assume the measurements are exact, then the circumference is three times the diameter: $\pi = 3$. But the passage is clearly not intended as a precise mathematical statement.

Page 90   Mission to Abisko.   *Mission to Abisko* (eds. J. Casti and A. Karlqvist), Perseus, New York 1999, 157–185.

Page 108   *Occasionally someone invents such a piece of machinery out of the blue, and proves all the experts wrong.*   A classic case is Louis De Branges's proof of the Bieberbach conjecture. See Ian Stewart, *From Here to Infinity*, Oxford University Press, Oxford 1996, 206.

Page 109   *Sir Peter Swinnerton-Dyer, has offered a simpler explanation of Fermat's claim.*   P. Swinnerton-Dyer, The justification of mathematical statements, *Philosophical Transactions of the Royal Society of London* Series A 363 (2005), 2437–2447.

Page 111   *Archimedes knew how to trisect an angle using a* marked *ruler and compass.*   Given angle AOB, draw BE parallel to OA, and the circle center B through O whose radius equals CD, where C and

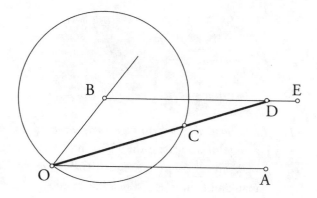

D are the marks on the ruler (thick line). Place the ruler so that it passes through O, while C lies on the circle and D lies on BE. Then angle AOC is one-third of angle AOB. See Underwood Dudley, *A Budget of Trisections*, Springer, New York 1987.

Page 111   *given a chessboard with two diagonally opposite corners missing, can you cover it with thirty-one dominoes?* The left figure shows the chessboard with its missing corners. The right figure shows a typical attempt to cover it: two squares are left uncovered.

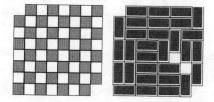

In contrast, if the two missing corners are adjacent to each other, then the puzzle is easily solved:

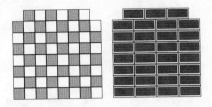

Page 113   *it is impossible to trisect the angle using an unmarked straightedge and compass.*   The first proof was given by Wantzel. See Ian Stewart, *Galois Theory*, Chapman and Hall /CRC, Boca Raton 2004.

Page 118   *a short calculation shows that with rare exceptions, the cubic equation associated with angle trisection is not like that.*   Ian Stewart, *Galois Theory*, Chapman and Hall /CRC, Boca Raton 2004.

Page 129   *Mathematicians are proud to trace their academic lineage through thesis advisers.*   There is a website dedicated to doing just that: http://www.genealogy. ams.org/

Page 130   *All of my Portuguese daughters have remained in mathematics.*   The story of the first, Isabel Labouriau, is one of the many fascinating biographical histories in a wonderful book about women in mathematics: *Complexities* (eds. Betty Anne Case and Anne M. Leggett), Princeton University Press, Princeton 2005.

Page 132   *a penetrating article in 1981 in* Mathematics Tomorrow.   Timothy Poston, Purity in applications, in *Mathematics Tomorrow* (ed. L. A. Steen), Springer, New York 1981, 49–54.

Page 133   *such gems as the law of quadratic reciprocity.*   This theorem, first proved by Gauss, states that if p and q are odd primes, then the equation $x^2 = mp + q$ has a solution in integers if and only if the related equation $y^2 = nq + p$ has a solution, except that if both p and q are of the form $4k + 3$, then one equation has a solution and the other does not. See G. A. Jones and J. M. Jones, *Elementary Number Theory*, Springer, London 1998.

Page 134   *the Titius–Bode law.*   This empirical pattern in the spacing of the planets was discovered by Johann Titius in 1766 and published by Johann Bode in 1772. Take the series 0, 3, 6, 12, 24, 48, 96, in which each number except the first is twice the preceding number. Add 4 to each term and divide by 10, to get 0.4, 0.7, 1.0, 1.6, 2.8, 5.2, 10.0. Omitting 2.8, these are very close to the distances from the sun to Mercury, Venus, Earth, Mars, Jupiter, and Saturn, respectively, measured in astronomical units. (By definition, the distance from Earth to the sun is one astronomical unit.) The asteroid Ceres neatly filled the gap at 2.8.

Page 136   *Karl Weierstrass found a simple continuous function that is differentiable* nowhere.   K. Falconer, *Fractal Geometry*, Wiley, New York 1990.

Page 139   *On the Enfeeblement of Mathematical Skills.*   J. Hammersley. On the enfeeblement of mathematical skills by "Modern Mathematics" and by similar soft intellectual trash in schools and universities, *Bulletin of the Institute of Mathematics and its Applications* 4 (1960) 66.

Page 142   *his "shape of a drum" paper is a gem.*   M. Kac. Can one hear the shape of a drum? *American Mathematical Monthly* 73 (1966) 1–23. Given the spectrum of tones that can be produced by a vibrating membrane in the plane, can you deduce its shape? Kac proved that its area and perimeter can be deduced. The general question was answered in the negative by C. Gordon, D. Webb, and S. Wolpert, One can't hear the shape of a drum, *Bulletin of the American Mathematical Society* 27 (1992) 134–138.

Page 142   *Hammersley's 2004 obituary in the* Independent on Friday.   *Independent on Friday*, 14 May 2004.

Page 144   *John Barrow argues the case like this.*   *Mission to Abisko* (eds. J. Casti and A. Karlqvist), Perseus, New York 1999, 3–12.

Page 151   The Man Who Knew Infinity.   Robert Kanigel, *The Man Who Knew Infinity*, Scribner's, New York 1991.

Page 173   *Grisha Perelman's alleged proof of the Poincaré conjecture.*   J. Milnor, Towards the Poincaré conjecture and the classification of 3-manifolds, *Notices of the American Mathematical Society* 50 (2003) 1226–1233, and M. T. Anderson, Geometrization of manifolds via the Ricci flow, *Notices of the American Mathematical Society* 51 (2004) 184–193.